STUDENT UNIT GUIDE

NEW EDITION

Edexcel AS Physics Unit 2
Physics at Work

Mike Benn

PHILIP ALLAN

Philip Allan, an imprint of Hodder Education, an Hachette UK company, Market Place, Deddington, Oxfordshire OX15 0SE

Orders
Bookpoint Ltd, 130 Milton Park, Abingdon, Oxfordshire OX14 4SB
tel: 01235 827827
fax: 01235 400401
e-mail: education@bookpoint.co.uk
Lines are open 9.00 a.m.–5.00 p.m., Monday to Saturday, with a 24-hour message answering service. You can also order through the Philip Allan website: www.philipallan.co.uk

© Mike Benn 2012

ISBN 978-1-4441-7146-4

First printed 2012
Impression number 5 4 3 2 1
Year 2017 2016 2015 2014 2013 2012

Cover photo: Fotolia

Printed in Dubai

Hachette UK's policy is to use papers that are natural, renewable and recyclable products and made from wood grown in sustainable forests. The logging and manufacturing processes are expected to conform to the environmental regulations of the country of origin.

P02082

Contents

Getting the most from this book

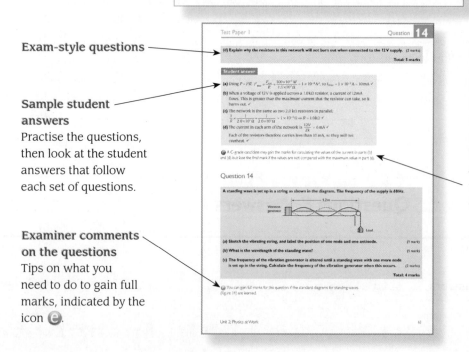

About this book

This guide is one of a series covering the Edexcel specification for AS and A2 physics. It offers advice for the effective study of Unit 2: Physics at Work. Its aim is to help you understand the physics — it is not intended as a shopping list that enables you to cram for the examination. The guide has two sections:

- **Content Guidance** — this section is not intended to be a detailed textbook. It offers guidance on the main areas of the content of Unit 2, with an emphasis on worked examples. These examples illustrate the types of question that you are likely to come across in the examination. The specification for Unit 2 states that, as well as the laws, theories and models of physics, practical applications of these should be explored. In this guide, practical applications will be mentioned where appropriate and may be referred to in questions, but you are expected to have encountered many examples already during your AS course.
- **Questions and Answers** — this comprises two unit tests, presented in a format close to that of an actual Edexcel examination and using questions similar to those in recent past papers, to give the widest possible coverage of the unit content. Answers are provided; in some cases, distinction is made between responses that might have been written by an A-grade candidate and those typical of a C-grade candidate. Common errors made by candidates are also highlighted so that you, hopefully, do not make the same mistakes.

A deep understanding of physics can develop only with experience, which means time spent thinking about physics, working with it and solving problems. This book provides you with a platform for doing this. If you try all the worked examples and the unit tests before looking at the answers, you will begin to think for yourself and develop the necessary techniques for answering examination questions effectively. In addition, you need to learn all the basic formulae, definitions and experiments. Thus prepared, you will be able to approach the examination with confidence.

The specification outlines the physics that will be examined in the unit tests and describes the format of those tests. This is not necessarily the same as what teachers might choose to teach or what you might choose to learn.

The purpose of this book is to help you with Unit Test 2, but don't forget that what you are doing is *learning physics*. The specification can be obtained from Edexcel, either as a printed document or from the web at **www.edexcel.com**.

Content Guidance

Waves

There are many types of wave. In Unit 2 you study the properties of **mechanical waves**, such as sound waves, and **electromagnetic waves**, with particular emphasis on the behaviour of light. Most types of waves transfer energy from the source to the observer (i.e. they are **progressive waves**). However, in some cases (**standing waves**) energy is not transmitted but rather changes form between potential energy and kinetic energy.

Wave terminology

The oscillating particles in a mechanical wave, e.g. in a stretched wire or slinky, can be described by a displacement–time graph for a single particle within the wave, or by the position of all the particles at a single instant along a section of the wave (displacement–distance graph — see Figure 1).

Examiner tip
These two types of graph are usually identical in shape. Therefore, always check the axes of a graph carefully so that you know exactly what is being represented.

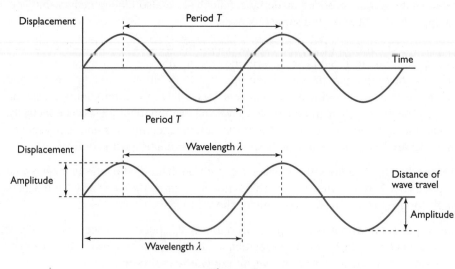

Figure 1

You need to learn the fundamental definitions that relate to wave motion. They are: **amplitude**, **period**, **frequency**, **wave speed** and **wavelength**. It is useful to refer to the above graphs to gain a full understanding of their meaning.

- **Amplitude**, y_0 — the maximum displacement of a particle from the midpoint of the oscillation. Unit: metre (m).
- **Period**, T — the time taken for one complete oscillation. Unit: seconds (s).
- **Frequency**, f — the number of oscillations per second. Unit: hertz (Hz).
- **Wave speed**, v — the distance travelled by the wave each second. Unit: metres per second (m s^{-1}).

- **Wavelength**, λ — the distance between consecutive points at which the oscillations are in phase. Unit: metre (m).

From these definitions, two useful equations can be established:

$f = \dfrac{1}{T}$, e.g. if there are 10 oscillations in 1 second, each one will have a period of 0.1 s.

$v = f\lambda$ This is often called the **wave equation**, and comes from the fact that a wave travels one wavelength in the time taken for one oscillation, so that:

$$v = \frac{\lambda}{T} = \frac{1}{T} \times \lambda = f\lambda$$

Examiner tip
It is worth noting that one complete cycle on a displacement–time graph represents one period, and one cycle of a displacement–distance graph covers a distance of one wavelength.

Worked example

The prongs of a tuning fork vibrate with a period of 2.5 ms. The speed of sound in air is 340 m s^{-1}. Calculate:

(a) the frequency of the emitted sound

(b) the wavelength of the sound in air

Answer

(a) $f = \dfrac{1}{T} = \dfrac{1}{2.5 \times 10^{-3} \text{ s}} = 400 \text{ Hz}$

(b) $\lambda = \dfrac{v}{f} = \dfrac{340 \text{ m s}^{-1}}{400 \text{ Hz}} = 0.85 \text{ m}$

Longitudinal and transverse waves

The difference between **longitudinal** and **transverse** waves can be illustrated using a slinky spring (Figure 2).

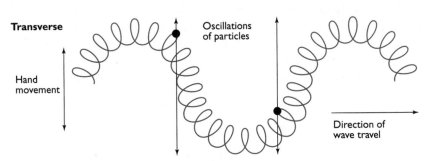

Figure 2

- In a **longitudinal** wave the particles oscillate back and forth along the line in which the wave progresses.
- In a **transverse** wave the particles oscillate at right angles to the direction of propagation of the wave.

Sound waves are examples of longitudinal waves: the back and forth motion of air molecules leads to alternate regions of high and low pressure, called **compressions** and **rarefactions** respectively. The oscillations of the particles can be observed by placing a lighted candle close to a large loudspeaker that is emitting a low-frequency sound (Figure 3).

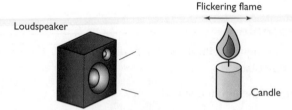

Figure 3

If a stone is thrown into a pond, a series of circular ripples can be seen moving outwards from the point where the stone entered the water. A duck floating on the water will 'bob' up and down in the same place as the wave passes along the pond surface (Figure 4). This demonstrates the transverse wave motion of the water particles.

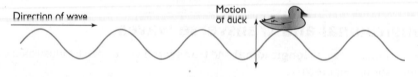

Figure 4

Other examples of transverse waves that you will encounter in Unit 2 are waves in stretched strings or wires and the variations of electric and magnetic fields in electromagnetic radiation.

Electromagnetic waves

Electromagnetic waves differ from the mechanical waves you have met so far, in that they do not consist of vibrating particles. These waves do not require a medium through which to travel, as they consist of regularly changing electric and magnetic fields. You do not need to know the properties of electric and magnetic fields for the AS examination (these will be covered in Unit 4 of the A2 course), but you will be required to have a general knowledge of the types of radiation that make up the electromagnetic spectrum, including similarities and differences between them, and to be able to describe some of their applications.

All electromagnetic radiation:

- travels in a vacuum with a speed of $3.0 \times 10^8 \, \text{m s}^{-1}$
- consists of oscillating electric and magnetic fields that are in phase and whose transverse variations lie within planes at right angles to each other

Examiner tip
Electromagnetic waves are effectively transverse waves, but be aware that there are no particles vibrating.

Table 1 gives some information on the production, properties and wavelengths of the radiation types that make up the electromagnetic spectrum.

Table 1

Type of wave	Wavelength range/m	Method of production	Properties and applications
γ-rays	10^{-16} to 10^{-11}	Excited nuclei falling to lower energy states	Highly penetrating rays. Used in medicine for destroying tumours, diagnostic imaging and sterilisation of instruments.
X-rays	10^{-14} to 10^{-10}	Fast electrons decelerating after striking a target	Similar to γ-rays, but the method of production means that their energy is more controllable. Used in medicine for diagnosis and therapy; in industry for detecting faults in metals and studying crystal structures.
Ultraviolet	10^{-10} to 10^{-8}	Electrons in atoms that were raised to high energy states by heat or by electric fields, falling to lower permitted energy levels	Used in fluorescent lamps and for detecting forged banknotes. Stimulates the production of vitamin D in the skin to cause tanning; makes some materials flouresce.
Visible light	4×10^{-7} to 7×10^{-7}		Light focused onto the retina of the eye creates a visual image in the brain. Can be detected by chemical changes to photographic film and electrical charges on the CCDs in digital cameras. Essential energy source for plants undergoing photosynthesis.
Infrared	10^{-7} to 10^{-3}		Radiated by warm bodies. Used for heating and cooking, and in thermal imaging devices.
Microwave	10^{-4} to 10^{-1}	High-frequency oscillators such as a magnetron; background radiation in space	Energy is transferred to water molecules in food by resonance at microwave frequencies. Used for mobile phone and satellite communications.
Radio	10^{-3} to 10^{5}	Tuned oscillators linked to an aerial	Wide range of frequencies allows many signals to be transmitted. Groups of very large radio-telescopes can detect extremely faint sources in space.

Examiner tip
You may be required to identify the nature of an electromagnetic wave having calculated its wavelength or frequency. You should be aware of the wavelength (and frequency) ranges for each section of the spectrum.

Knowledge check 1

State the regions of the electromagnetic spectrum in which the frequency of the radiation is: (a) 10 GHz, (b) 3×10^{16} Hz, (c) 98 MHz.

Knowledge check 2

State the regions of the electromagnetic spectrum in which the wavelength of the radiation is: (a) 500 nm, (b) 900 nm, (c) 3.0×10^{-12} m.

Worked example

A transmitter sends out electromagnetic waves of frequency 9.3 GHz. Calculate the wavelength of these waves, and state the region of the electromagnetic spectrum to which they belong.

Answer

Using $v = f\lambda$, we get:

$$\lambda = \frac{v}{f} = \frac{3.0 \times 10^8 \, \text{m s}^{-1}}{9.3 \times 10^9 \, \text{Hz}} = 3.2 \times 10^{-2} \, \text{m}$$

These waves are part of the microwave section of the spectrum.

Superposition and interference

Earlier in this section, displacement–distance graphs and a displacement–time graph for a particle oscillating within a wave were used to illustrate some basic wave definitions. Such graphs can also be used to explain the ideas of **phase** and **phase difference**, which will help us understand the effects of superposition and interference that occur when similar waves meet at a point. Consider the displacement–time graph in Figure 5.

Examiner tip

The displacement–time graph for a longitudinal wave is identical to that of a transverse wave, even though the particles are oscillating in a different plane.

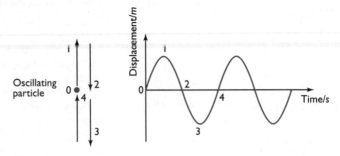

Figure 5

Suppose that the timing of the oscillation begins when the particle is moving upward through the midpoint; this is the point marked 0 on the graph. The points labelled 1, 2, 3 and 4 represent the stages, or **phases**, of one complete cycle of the vibration.

Point 1 — the particle is at the positive extreme position. It has completed one quarter of a cycle and is said to be 90° or $\frac{\pi}{2}$ radians **out of phase** with point 0.

Point 2 — the particle is moving down through the midpoint. It is exactly half a cycle behind point 0, which corresponds to a phase difference of 180° or π radians. Points 0 and 2 are said to be **in antiphase**.

Point 3 — the particle is at the negative extreme position, three-quarters of a cycle from the starting point; this corresponds to a phase difference of 270° or $\frac{3\pi}{2}$ radians.

Point 4 — the particle has reached the end of one cycle and is at the same position and moving in the same direction as point 0. The phase difference between point 4 and point 0 is 360° or 2π radians, and we say that the points are now back **in phase**.

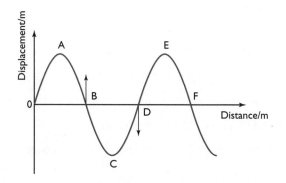

Figure 6

From the displacement–distance graph in Figure 6, it can be seen that the phase of particles in the wave varies continuously along one complete wavelength. Hence the motion of particle A is in phase with that of particle E, but is π radians out of phase (i.e. in antiphase) with the motion of particle C. Similarly, B is in phase with F but in antiphase with D.

Worked example

A wave motion is described by the graphs in Figure 7.

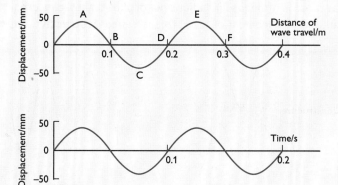

Figure 7

(a) State the amplitude of the wave.

(b) State the wavelength.

(c) Give all the pairs of points, other than {0 and D} and {A and E}, which are in phase.

(d) Give all the pairs of points, other than {0 and B} and {A and C}, which are in antiphase.

(e) Give all the pairs of points, other than {0 and A} and {A and B}, that have a phase difference of $\frac{\pi}{2}$.

(f) State the period of the motion.

(g) Determine the frequency of the motion.

(h) Hence calculate the wave speed.

Answer

(a) Amplitude = maximum displacement = 40 mm

(b) Wavelength λ = distance between 0 and D = 0.2 m

(c) B and F are the only other points that are in phase.

(d) {B and D}, {C and E}, and {D and F} are in antiphase.

(e) {B and C}, {C and D}, {D and E}, and {E and F} differ in phase by $\frac{\pi}{2}$.

(f) Period T = time for one oscillation = 0.1 s

(g) Frequency $f = \dfrac{1}{T} = \dfrac{1}{0.10\,s} = 10\,Hz$

(h) Wave speed $v = f\lambda = 10\,s^{-1} \times 0.2\,m = 2.0\,m\,s^{-1}$

Superposition

When two waves of the same type (e.g. two water waves) arrive at the same point, their displacements will combine to result in a wave of different amplitude.

If the waves are in phase, they add together to give a wave whose amplitude is the sum of the amplitudes of the original waves. This is called **constructive superposition**.

If the two waves are in antiphase (π radians out of phase), they 'cancel each other out', giving a resultant displacement of zero. This is called **destructive superposition** (Figure 8).

Examiner tip
Superposition only occurs when two waves *of the same type* meet at a point

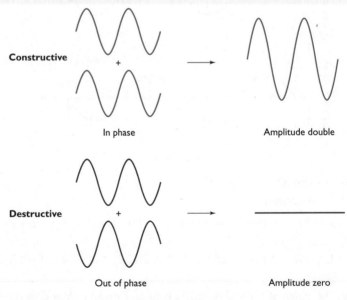

Figure 8

Wavefronts

You will probably have seen a ripple tank being used to demonstrate the properties of waves on a water surface (if not, there are plenty of simulations online that can be accessed by entering 'ripple tank' into a search engine). Waves can be observed moving across the surface as a series of lines (Figure 9).

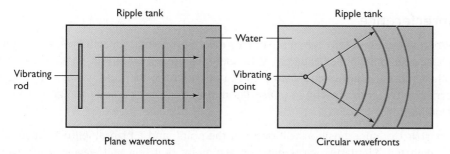

Figure 9

The darker lines represent the 'crests' of the wave and the light regions are at the 'troughs', where the water is shallower. The particles in the wave along a particular crest will all be in phase; such a line is known as a **wavefront**.

As a wave progresses, it can be seen that the wavefronts are always at right angles to the direction in which the wave is moving. Therefore, for waves radiating out from a point source the wavefronts will be circular (or spherical in three dimensions), whereas plane wavefronts represent waves moving along parallel paths.

The distance between consecutive wavefronts is the wavelength of the wave in the medium through which it is travelling.

Coherence

Wave sources are said to be **coherent** if:
- the waves are of the same type
- the waves all have the same frequency
- the sources are always in phase or maintain a constant phase difference

Coherent sources can be obtained from the same wavefronts or by connecting transmitters to the same oscillator (Figure 10).

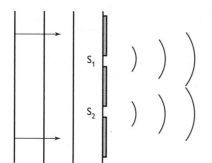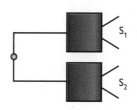

Figure 10

Conventional light sources, such as tungsten filament lamps, cannot be used as coherent sources because they emit photons of radiation (see the 'Nature of light' section) with different wavelengths and random phase differences.

> A **wavefront** is a line joining points in a wave that are in phase with each other.

> **Examiner tip**
> When drawing wavefronts in examinations, you must ensure that, if the wavelength is constant, the separations of the lines must all be the same.

> **Examiner tip**
> A common error in examinations is stating simply that 'coherent sources are in phase' — this could be true for any pair of sources at a certain instant but does not suffice to define coherence. The idea of a phase relationship between the sources, which remains constant for all times, must be stressed.

Lasers give out groups of photons which have the same frequency and are all in phase, so the emitted radiation is generally considered coherent.

Interference

When waves from two coherent sources meet at a point, they will have a constant phase relationship. For example, if the sources are constantly in phase with each other and the point is equidistant from both sources, then the waves will always arrive at the point exactly in phase (assuming that they passed through the same medium).

Constructive superposition will occur at the point, and a high-amplitude wave will be detected. Similarly, waves meeting at a point that is a half-wavelength further away from one source than the other will always be out of phase at that point, so destructive superposition results.

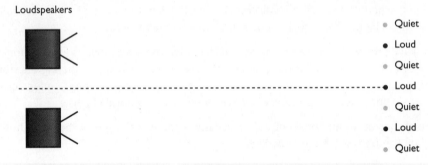

Figure 11

If two loudspeakers are connected to the same output from a signal generator and placed about 1 metre apart in a large hall (Figure 11), a listener walking across the hall will detect a series of loud and quiet regions at regular intervals. This pattern of alternating maximum and minimum intensities is called an **interference pattern**.

Any pair of coherent sources can generate such interference patterns, including sources of water, radio and light waves (Figure 12).

Examiner tip

Note that superposition of waves does not necessarily lead to interference. Waves can meet randomly at a point and momentarily cause an increase or decrease in amplitude, but for a fixed pattern to occur, the superposition needs to be continuous at that point.

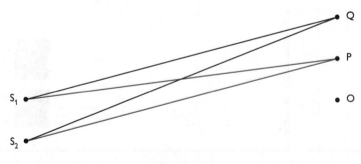

Figure 12

If the coherent sources S_1 and S_2 are in phase, the waves meeting at O will have travelled the same distance and so will be in phase and interfere constructively. Thus there will always be a wave of maximum intensity at this point.

At P, the wave from S_2 has travelled half a wavelength farther than the distance travelled by the wave from S_1, so the waves interfere destructively at this point.

At Q, the distance from S_2 is a full wavelength greater than that from S_1, so, like at O, constructive interference takes place.

The difference between the distances from the two sources to a given point, $|S_2P - S_1P|$, is called the **path difference**.

For an interference pattern to be observed:
- two coherent sources are required
- the sources should be of similar amplitude

If the sources are in phase, then:
- when the path difference is zero or equal to a whole number of wavelengths, constructive interference takes place
- when the path difference is an odd number of half-wavelengths, destructive interference occurs

Knowledge check 3

State the conditions required for an interference pattern to be formed.

Worked example

Two speakers are placed about 1 metre apart in a school hall. They are connected to the same signal generator so that the sounds from each are emitted in phase. A student standing at the opposite end of the hall at a point equidistant from both speakers hears a loud sound from the speakers. The student then walks across the hall parallel to the line from speaker to speaker, and hears the sound intensity decrease and increase alternately. She makes a mark on the floor at the position of the third intensity minimum from her starting point. The distance of this point is measured to be 6.20 m from one speaker and 8.20 m from the other.

(a) Calculate the wavelength of the sound in air.

(b) If the frequency of the signal generator is 420 Hz, determine the speed of sound in air.

(c) If the frequency is increased, explain how the pattern of maxima and minima will be affected.

Answer

(a) Path difference = 8.20 m – 6.20 m = 2.00 m

For the third minimum, path difference is 2.5λ, so $\lambda = \dfrac{2.00\,\text{m}}{2.5} = 0.80\,\text{m}$

(b) $v = f\lambda = 420\,\text{Hz} \times 0.80\,\text{m} = 336\,\text{m s}^{-1}$

(c) Increasing the frequency reduces the wavelength. The path difference that determines the positions of maxima and minima will therefore decrease, so the loud and quiet positions will become closer together.

Applications of interference

In addition to finding frequency, wavelength and speed for a variety of waves, interference effects are used extensively in the optics industry. Lens surfaces can be

ground to tolerances of around 500 nm with the aid of optical interferometers, which make use of small changes in the interference patterns produced by reflections of light from glass surfaces when the separation between the surfaces is altered.

Active noise control is now common for pilots and tractor drivers. The noise is recorded and electronically adjusted before being fed back into earpieces in such a way that it is out of phase with the original sound, leading to destructive interference. The noise is effectively cancelled out.

Standing waves

Standing (stationary) waves are a particular example of interference: they are set up when two waves of equal frequency and amplitude travelling at the same speed in opposite directions superimpose.

Consider two coherent sources that are facing one another, an integral number of half-wavelengths apart (Figure 13). The same principles can be applied as for the production of an interference pattern described earlier.

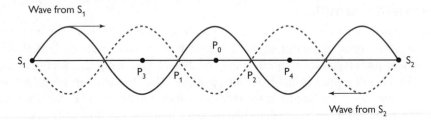

Figure 13

At the midpoint P_0 between S_1 and S_2, the path difference from the sources is zero, so the waves will always be in phase and superimpose constructively. At this position, the particles oscillate with maximum amplitude. However, at P_1 and P_2, the path difference is half a wavelength, so destructive interference takes place and the particles at these positions always have zero amplitude. In other words, there is no movement of particles at P_1 or P_2; such a point is termed a **node**.

At points P_3 and P_4 the path difference equals one whole wavelength. These, like P_0, are therefore positions where the wave has maximum amplitude, called **antinodes** (Figure 14).

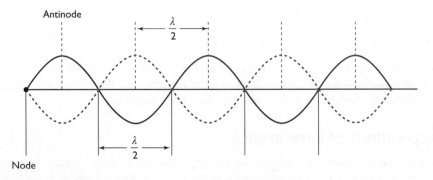

Figure 14

From Figure 14 you can see that the distance between adjacent nodes (and between adjacent antinodes) is half a wavelength.

A progressive wave moves by virtue of the phase difference between adjacent particles. In a standing wave, however, all the points between any two consecutive nodes are in phase at all times. Progressive waves transfer energy along the direction of wave travel, but in stationary waves energy changes form between kinetic energy and potential energy as the particles oscillate about their mean positions (except at the nodes, where the energy is zero).

The differences between progressive and stationary waves are summarised in Table 2.

Examiner tip
Never refer to points of maximum and minimum amplitude on a progressive wave as antinodes and nodes. The terms only apply to stationary waves.

Table 2

Progressive waves	Stationary waves
Energy is transferred in the direction of wave travel.	Energy is stored within each vibrating particle.
All points along the wave have the same amplitude.	Amplitude varies between a maximum value at the antinodes to zero at the nodes.
Adjacent points in the wave have a different phase relationship.	All points between each pair of consecutive nodes have a constant phase relationship.

Worked example

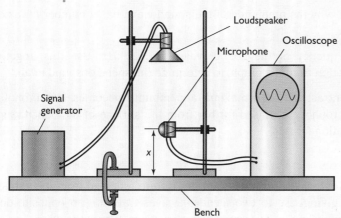

Figure 15

Figure 15 shows an experiment with sound waves. The height of the trace on the oscilloscope is proportional to the amplitude of the sound wave at the microphone. When the vertical distance, x, between the microphone and the bench is varied, the amplitude of the sound waves is found to vary as shown in Figure 16.

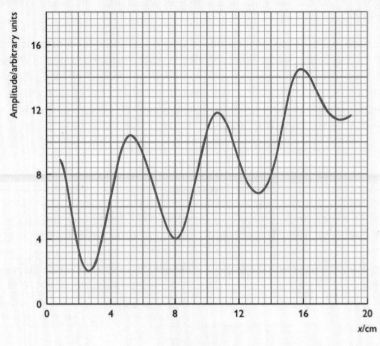

Figure 16

(a) Explain why the amplitude of the sound wave has a number of maxima and minima.

(b) The frequency of the sound waves is 3.20 kHz. Use this, together with information from the graph, to determine the speed of sound in air.

(c) The contrast between maxima and minima becomes less pronounced as the microphone is raised further from the surface of the bench. Suggest an explanation for this.

Answer

(a) The sound waves from the loudspeaker are reflected from the surface of the bench, giving rise to two identical waves travelling in opposite directions, which superimpose to form a stationary wave. Where the waves are in phase, an antinode of maximum intensity is formed; where the waves are in antiphase, a node of minimum intensity is formed.

(b) From the graph, the distance between the first and fourth minima (nodes) is
18.2 cm − 2.6 cm = 15.6 cm

As the distance between adjacent nodes is $\lambda/2$ we have that:

$$3 \times \frac{\lambda}{2} = 15.6\,\text{cm}$$

$$\lambda = \frac{15.6\,\text{cm} \times 2}{3} = 10.4\,\text{cm}$$

so $v = f\lambda = (3.20 \times 10^3\,\text{s}^{-1}) \times (10.4 \times 10^{-2})\,\text{m} = 333\,\text{m s}^{-1}$

(c) An ideal stationary wave, with all maxima and minima having the same amplitude, is formed from two travelling waves (in opposite directions) of the same amplitude. However, in this experimental situation, the reflected wave loses some energy upon reflection, and its amplitude decreases even further as it moves towards the loudspeaker. In contrast, the amplitude of the transmitted wave is greatest near the loudspeaker. Consequently, the amplitudes of the transmitted and reflected waves become less similar as the microphone moves towards the speaker, causing the maxima and minima to be less pronounced.

Applications of standing waves

As shown in the above example, measurements of wavelengths from the separations between nodes or antinodes can be used, via the $v = f\lambda$ equation, to determine frequency or wave speed. The example also demonstrates how interference of an incident and a reflected wave can create a standing wave.

Knowledge check 5

Calculate the wavelength of the sound from a trumpet, if its frequency is 256 Hz and the speed of sound in air is 340 m s^{-1}.

This is relevant to understanding the production of sound by stringed and piped musical instruments. When a guitar string is plucked, the disturbance travels to the ends of the string where it gets reflected. The reflections travel in opposite directions and superimpose to produce a standing wave in the string. Some of the energy in the string is transferred to the surrounding air, generating a sound of the same frequency as that of the string's vibrations.

A number of different standing waves are produced in the string. Because each end of the string is fixed, these ends must be nodes. The standing wave with the longest possible wavelength is the one which has just a single antinode at the midpoint of the string; this wavelength corresponds to the **fundamental frequency** for the string. Other standing waves, with more nodes (N) and antinodes (A), can be set up as shown in Figure 17.

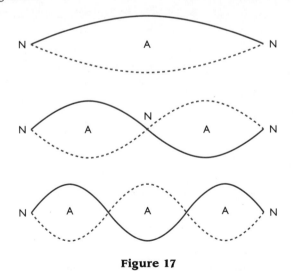

Figure 17

The standing waves with shorter wavelengths are called **overtones**, or **harmonics** if they have a frequency that is an integer multiple of the fundamental frequency.

The frequency of the sound emitted by a stringed instrument is inversely proportional to the length of the string. It is also dependent on the speed of the waves in the string. The wave speed is affected by the mass per unit length of the string and the tension applied across its ends.

Wind instruments also produce stationary waves. In a recorder, for example, there will be an antinode at each end; the fundamental frequency can be varied (generating different musical notes) by opening or closing the stops with your fingers (Figure 18).

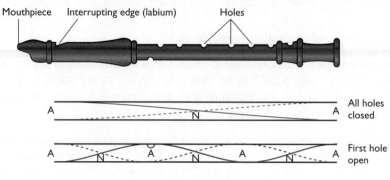

Figure 18

The air temperature can have an effect on the notes emitted from a wind instrument. As the temperature rises, the speed of sound in air increases; so from the wave equation $v = f\lambda$ we can see that, for a standing wave of fixed wavelength, the frequency will increase, resulting in a higher-pitched note.

Summary

After studying this section, you should be able to:
- define the terms amplitude, period, frequency, wave speed and wavelength of a wave, and use the wave equation $v = f\lambda$
- explain the difference between transverse and longitudinal waves
- recall the regions of the electromagnetic spectrum and be aware of the different properties, wavelengths and frequencies of each type of radiation
- explain how two similar waves can be superimposed to produce a wave of greater or smaller amplitude
- define coherence and explain how two coherent sources can produce an interference pattern
- describe how stationary waves are formed and determine the frequency or wavelength of such waves in strings and pipes

Refraction

Waves travel at different speeds in different media. For example, the speed of sound is approximately $340\,\text{m s}^{-1}$ in air at $20°C$, $1500\,\text{m s}^{-1}$ in water and $5000\,\text{m s}^{-1}$ in steel; light travels at $3.0 \times 10^8\,\text{m s}^{-1}$ in a vacuum, $2.0 \times 10^8\,\text{m s}^{-1}$ in glass and $2.3 \times 10^8\,\text{m s}^{-1}$ in water.

When a wave is incident at an interface between two transmitting media, the change in speed can result in a change in the direction of travel of the wave. This effect is known as **refraction** (Figure 19).

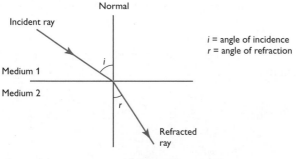

Figure 19

If the wave slows down as it passes from one medium into the next, it will deviate towards the normal, as shown in the diagram above. If the wave speeds up as it enters the second medium, it will deviate away from the normal.

The ratio of the wave's speed in the first medium to its speed in the second medium, denoted by $_1\mu_2$, is called the **refractive index** for that transmission:

$$_1\mu_2 = \frac{\text{speed in medium 1}}{\text{speed in medium 2}} = \frac{v_1}{v_2}$$

The refractive index can be expressed in terms of the angles of incidence and refraction. This relationship is known as **Snell's law**.

$$_1\mu_2 = \frac{\sin i}{\sin r}$$

where i is the angle of incidence and r is the angle of refraction.

Worked example

A beam of ultrasound is passed through a fine membrane that separates air and carbon dioxide. The beam is incident in the air at an angle of 55° to the normal drawn at the interface and enters the carbon dioxide at an angle of 46° to the normal.

(a) Calculate the refractive index for sound travelling from air into carbon dioxide.

(b) Determine the speed of sound in carbon dioxide, given that the speed in air is 340 m s⁻¹.

Answer

(a) By Snell's law, $\mu = \dfrac{\sin 55°}{\sin 46°} = 1.14$

(b) Let v be the speed of sound in carbon dioxide.

Then $1.14 = \dfrac{340\,\text{m s}^{-1}}{v}$ so $v = \dfrac{340\,\text{m s}^{-1}}{1.14} = 300\,\text{m s}^{-1}$ (to 2 s.f.)

Refraction of light

In many situations, light is refracted when it passes from air into glass, water or some other transparent medium. It is often useful to specify the **absolute refractive index** of a material for such refractions.

$$\mu_m = \frac{\text{speed of light in a vacuum}}{\text{speed of light in the medium}}$$

Examiner tip
Refraction occurs for all types of wave when the wave speed changes at a boundary between different media.

The **refractive index** for a wave moving from one medium to another is the ratio of the wave speed in the first medium to that in the second.

Snell's law states that the refractive index for a wave travelling from one medium to another is given by the ratio of the sine of the angle of incidence to the sine of the angle of refraction.

Examiner tip
The property of a material that relates to the wave speed through it is often referred to as the 'optical density'. The more optically dense the medium is, the slower the wave speed. So a wave travelling into a medium of greater optical density will slow down and bend towards the normal.

The **absolute refractive index** of a material is the ratio of the speed of light in a vacuum to its speed in the material.

Examiner tip
The speed of light cannot exceed $3.0 \times 10^8\,\mathrm{m\,s^{-1}}$ (the value in a vacuum). It follows that the refractive index for light passing through any medium must be greater than 1. If you calculate a value that is less than 1, it is likely that you have used the wrong angles.

A useful way of expressing Snell's law in terms of absolute refractive indices is:

$$\mu_1 \sin \theta_1 = \mu_2 \sin \theta_2$$

where θ_1 is the angle of incidence in the first medium and θ_2 is the angle of refraction in the second medium.

Worked example

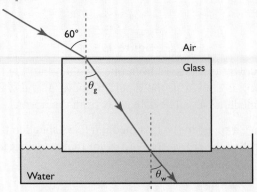

Figure 20

A ray of light enters a glass block at an angle of incidence of 60° and emerges through a layer of water on the opposite side (Figure 20). If the refractive index of glass is 1.55 and that of water is 1.33, calculate:

(a) the angle of refraction in the glass, θ_g

(b) the angle of refraction in the water, θ_w

Answer

(a) For the ray entering the glass block from above, we have $1.55 = \dfrac{\sin 60°}{\sin \theta_g}$ from Snell's law.

This gives $\sin \theta_g = \dfrac{\sin 60°}{1.55}$ and hence $\theta_g = 34°$

(b) For the ray passing from the glass block into the water underneath, we have:

$$\mu_g \sin \theta_g = \mu_w \sin \theta_w$$

$$1.55 \sin 34° = 1.33 \sin \theta_w$$

giving $\theta_w = 41°$

Total internal reflection

When light travels from an optically dense (low speed) medium to a less dense (higher speed) medium, such as from glass into air, it deviates away from the normal (Figure 21).

For small angles of incidence θ_1, most of the light will be refracted out of the glass block. As θ_1 is increased, the refracted ray will deviate further from the normal until it is eventually at 90° to the normal (i.e. θ_2 becomes 90°), parallel to the interface.

For even larger angles of incidence, no light will be able to leave the glass, and **total internal reflection** takes place. The angle at which this takes place is called the critical angle.

The **critical angle**, C, is the angle of incidence for which total internal reflection just takes place, i.e. it is the angle that leads to the light being refracted by exactly 90°.

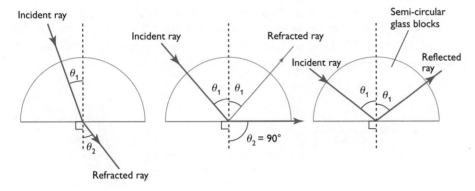

Figure 21

Applying Snell's law in this situation, we get:

$$\mu_1 \sin C = \mu_2 \sin 90°$$

$$\sin C = \frac{\mu_2}{\mu_1}$$

If the second medium is air, then μ_2 is approximately 1, so $\sin C = \frac{1}{\mu_1}$.

Worked example

The refractive index of glass is 1.50 and that of water is 1.33. Calculate the critical angle for light passing from:

(a) glass to air

(b) water to air

(c) glass to water

Answer

(a) $\sin C = \frac{1}{\mu_g} = \frac{1}{1.50}$, $C = 42°$

(b) $\sin C = \frac{1}{\mu_w} = \frac{1}{1.33}$, $C = 49°$

(c) $\sin C = \frac{\mu_w}{\mu_g} = \frac{1.33}{1.50}$, $C = 62°$

Applications of total internal reflection

Reflections from silvered glass mirrors occur at both faces, producing a blurred image which is unsuitable for precision optical instruments. For reflections in binoculars, reflex cameras and the laser beam splitters of CD players, the inner surfaces of 45° prisms are used (Figure 22).

In the food industry, the sugar concentrations of liquids can be measured from the refractive index of the solutions.

The most important recent developments in communications have probably been in the area of fibre optics. Glass fibres as thin as human hair have a core surrounded by a cladding of lower refractive index. This ensures that light passing through the core will be totally internally reflected as long as it impinges on the core–cladding boundary at an angle greater than the critical angle (Figure 22).

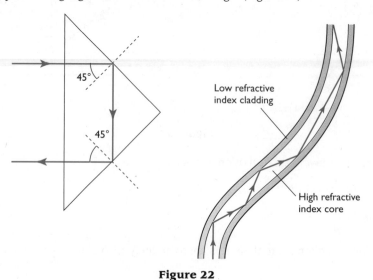

Figure 22

Examiner tip

Remember that total internal reflection only occurs at dense to less dense boundaries (high refractive index to low refractive index).

Worked example

A ray of light is directed at the midpoint of a semi-circular glass block. The angle of incidence is adjusted until total internal reflection just takes place. The critical angle is measured to be 41°.

The block is now placed onto a glass slide, with a layer of sugar solution between the block and the slide. The critical angle is now found to be 67°.

Calculate:

(a) the refractive index of the glass

(b) the refractive index of the sugar solution

Knowledge check 6

Calculate the critical angle for light transmitted through an optical fibre, if the refractive index of the core is 1.56 and that of the cladding is 1.48.

Answer

(a) $\sin 41° = \dfrac{1}{\mu_g}$, so $\mu_g = \dfrac{1}{\sin 41°} = 1.52$

(b) $\sin 67° = \dfrac{\mu_s}{\mu_g} = \dfrac{\mu_s}{1.52}$, hence $\mu_s = 1.52 \sin 67° = 1.40$

Plane polarisation

In normal light waves, the electric vector oscillates in all directions within a plane perpendicular to the direction of wave travel. If such light is passed through a sheet of Polaroid®, the oscillations in all directions but one (i.e. the direction of the transmission axis) will be absorbed. The light emerging from the sheet has its electric vector oscillating in one direction only and is said to be **plane polarised**.

Suppose that a second sheet of Polaroid® (or **analyser**) is held beyond the first sheet (the **polariser**). As this analyser is rotated, alternating maximum and minimum (virtually zero) light intensities will be observed every 90°, depending on whether the transmission axis of the analyser becomes parallel or perpendicular to the transmission axis of the polariser. These phenomena are illustrated in Figure 23.

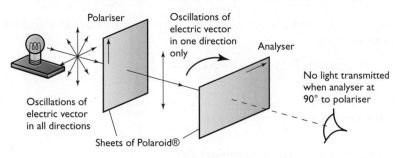

Figure 23

Microwaves can also be used to demonstrate polarisation. Microwave generators of the type found in school laboratories produce plane polarised waves of wavelength about 3 cm. The plane polarisation can be checked by using a suitable detector, such as an aerial, connected to an amplifier and an ammeter. When the aerial is parallel to the plane of polarisation, the maximum signal will be received. As the aerial is rotated about an axis joining it and the microwave transmitter, the signal intensity will diminish, reaching a minimum when the aerial has rotated through 90°.

Worked example

A microwave generator produces plane polarised waves of wavelength 30 mm.

(a) What is meant by a *plane polarised* electromagnetic wave?

(b) Draw a labelled diagram of the apparatus you would use to demonstrate that the microwaves are plane polarised.

(c) What is the frequency of these microwaves?

(d) What does this experiment show you about the nature of electromagnetic waves?

Answer

(a) A plane polarised electromagnetic wave is one in which the oscillations of the electric vector are in one direction only, in a plane perpendicular to the direction of wave travel.

Plane polarised waves are such that the particles, or fields, always oscillate in the same plane.

> **Examiner tip**
> Longitudinal waves cannot be plane polarised because the oscillations are always parallel to the direction of wave motion.

(b)

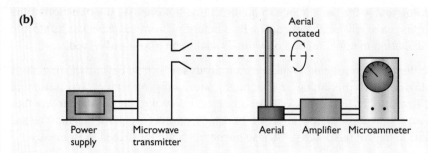

(c) Using $c = f\lambda$, where $c = 3.0 \times 10^8\,\mathrm{m\,s^{-1}}$ (as microwaves are electromagnetic waves), we have:

$$f = \frac{c}{\lambda} = \frac{3.0 \times 10^8\,\mathrm{m\,s^{-1}}}{30 \times 10^{-3}\,\mathrm{m}} = 1.0 \times 10^{10}\,\mathrm{Hz}$$

(d) As the microwaves are plane polarised, this experiment shows that electromagnetic waves are transverse.

Applications of plane polarised light

Some complex molecules can rotate the plane of polarisation of the transmitted light. In sugar solutions, the angle of rotation is dependent on the concentration of the solution.

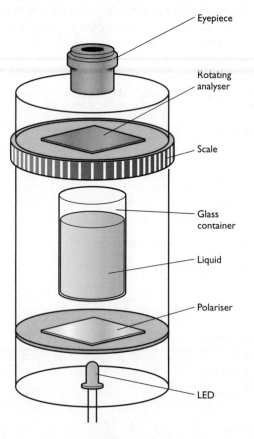

Figure 24

With distilled water in the container shown in Figure 24, the analyser is rotated until its transmission axis is at right angles to that of the polariser, thus blocking out the light. A sugar solution is then placed in the container. Because the solution rotates the plane of polarisation, light will again be able to emerge. The degree of rotation of the plane of polarisation is measured by adjusting the analyser until the light disappears once more.

Perspex models of mechanical components are tested using stress analysis. When the model is loaded and viewed with light passed through a polariser and an analyser, a multicoloured stress pattern is revealed, which is due to rotation of the plane of polarisation by the strained regions. Potential areas of weakness can be detected from these observations.

Liquid crystal displays (LCDs) work by using electric fields to align crystals that rotate the plane of polarisation of light.

Diffraction

Diffraction occurs when a wavefront is disturbed by an obstacle or passes through an aperture. This is best illustrated by the water waves on a ripple tank (Figure 25).

> **Diffraction** is the spreading of waves after they have passed through an aperture, or been disturbed by an obstacle.

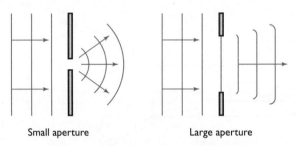

Small aperture   Large aperture

Figure 25

When the wavelength of the wave is of the same order of magnitude as the width of the gap, a wavefront spreads out in an almost (semi-)circular shape after passing through the gap, as in the left-hand diagram in Figure 25.

This explains why, even when someone is out of sight behind an open door, you can still hear the person talking — the wavelength of sound is about the same as the width of the door. Light, however, has a wavelength on the order of 500 nm, much narrower than the door. Therefore it will undergo no discernable diffraction through the doorway.

For light, diffraction effects can be observed when the aperture is very small. If a laser beam is directed at the gap between the jaws of vernier callipers that are almost closed, a pattern can be seen on a screen beyond the gap. In addition to a central band of maximum intensity, there will be a series of bright and dark lines that arise from the effects of interference between the disturbed wavefronts (Figure 26). If the gap is narrowed, the central maximum will spread outwards.

> **Knowledge check 8**
>
> Explain why in some places in hilly regions a radio signal may be received, but a mobile phone is unable to pick up a signal from a transmitting dish that is close to the radio transmitter.

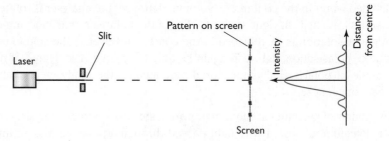

Figure 26

Diffraction patterns can also be observed when bright lights are shone through fine materials such as net curtains.

X-ray diffraction patterns, created by passing X-rays through regular crystal structures, were used to examine the atomic structures of many elements during the compilation of the periodic table. A similar method was used to demonstrate the wave–particle nature of the electron (Figure 27).

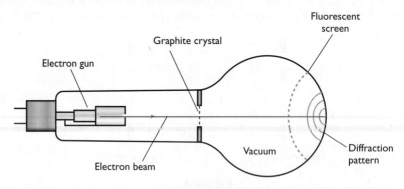

Figure 27

Atomic separations in the graphite crystal are in the order of 10^{-11} m, and electron diffraction suggests that the wavelength must be of the same order of magnitude. If the voltage across the tube is increased, giving the electrons more energy, then the rings of the diffraction pattern contract inwards, indicating a reduction in wavelength. The relationship between wavelength and energy of electromagnetic radiation will be explored in more detail in the 'Nature of light' section.

Reflection

The laws of reflection for light rays state that:
- the angle of incidence equals the angle of reflection (Figure 28)
- the incident ray, the reflected ray and the normal at the point of incidence all lie in the same plane

Unit 2 puts more emphasis on normal reflections, i.e. reflections at right angles to the surface, occurring at interfaces between different media.

For light that is incident at right angles to a glass surface, some of it will be reflected from the surface while the rest will be transmitted (refracted) through the glass. The fraction of light that gets reflected depends on the relative refractive index going from air to glass.

For sound, the proportion that gets reflected from the interface between two media depends upon a property known as the **acoustic impedance**, which is related to the speed of sound in a medium and the medium's density.

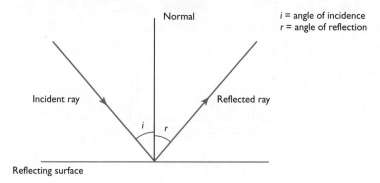

Figure 28

Pulse-echo techniques

The pulse-echo technique is a method for measuring the speed of a wave, or the distance from a reflecting surface, by finding the time taken for a short pulse of radiation to reflect back to the original position. The method is commonly used to determine the speed of sound in air (Figure 29).

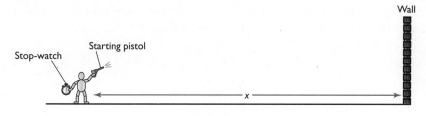

Figure 29

The operator starts the stop-watch when the gun is fired, and then stops the watch when the echo is heard, recording the time t. The speed of sound is found as follows:

$$\text{speed} = \frac{\text{distance}}{\text{time}} = \frac{2x}{t}$$

If we take the speed v as known, then the distance x can be calculated as:

$$x = \frac{vt}{2}$$

It is important to stress that, for this technique to work, the duration of the 'pulse' (the bang of the gun in this case) needs to be much shorter than the time taken for the wave to return. It would not be possible to time an echo from a source that emits a continuous sound.

In medicine, A-scans using ultrasound with frequencies of the order of 1 MHz can determine the depth of structures within the human body (Figure 30).

Examiner tip

It is a common error for candidates to forget about the return distance of the echo.

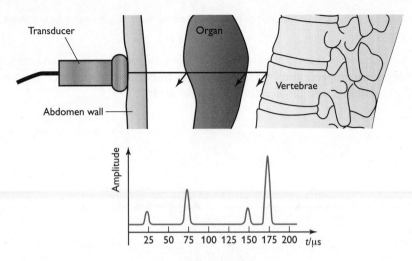

Figure 30

A pulse of ultrasound enters the body from the transducer. At each interface between different tissues in the body, a fraction of the sound gets reflected. These echoes are displayed on an oscilloscope that shows the time taken for each reflection to return to the transducer. The amplitude of each detected echo-signal represents the relative amount of ultrasound reflected at the corresponding interface.

Worked example

Use the timescale of the oscilloscope graph in Figure 30 to determine the distance of the organ from the inner abdomen wall and the thickness of the organ. The speed of sound is $1500\,\text{m}\,\text{s}^{-1}$ in soft tissue and $1560\,\text{m}\,\text{s}^{-1}$ in the organ.

Answer

Using $x = \frac{vt}{2}$, from the wall to the organ we have:

$$x = \frac{1500\,\text{m}\,\text{s}^{-1} \times (75-25) \times 10^{-6}\,\text{s}}{2} = 3.8 \times 10^{-2}\,\text{m}$$

From the front to the back of the organ:

$$x = \frac{1560\,\text{m}\,\text{s}^{-1} \times (150-75) \times 10^{-6}\,\text{s}}{2} = 5.9 \times 10^{-2}\,\text{m}$$

Knowledge check 9

Why must *pulses* of ultrasound be used in medical imaging?

Doppler effect

When a racing car moves past, you will hear the pitch of the engine noise shift from high to low. This is known as the **Doppler effect**, and can be observed whenever there is relative motion between the wave source and the observer (Figure 31).

As the source moves towards the observer, the wavefronts are effectively 'squeezed' together; thus, as perceived by the observer, the wavelength is reduced and so the frequency increases.

As the source moves away from the observer, the wavelength is effectively increased, causing a reduction in the observed frequency.

The **Doppler effect** is frequency change observed when the source of the wave is moving relative to the observer.

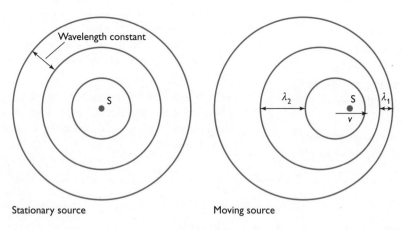

Figure 31

If the wave speed and frequency of the source are known, then by measuring the change in frequency, the relative speed between source and observer can be calculated.

In Unit 5 of the A2 course you will study the theory of the expanding universe. Observations of spectral lines from distant stars show that their frequency is less than what it would be for the same lines produced by stationary sources. The spectra have been displaced towards the red (longer-wavelength) end of the visible spectrum, therefore the effect is called the 'red shift'.

The Doppler effect can be used in conjunction with pulse-echo techniques, e.g. in ultrasound probes, for certain measurements in medicine. For example, to find the rate of blood flow in an artery, an ultrasound beam is reflected from the particles in the bloodstream back to the transducer; by measuring the change in frequency of the reflected beam, the rate of flow can be calculated. (A similar principle underlies the 'radar traps' used to measure the speed of motor vehicles.)

Another application of the Doppler effect is in measuring fetal heart rates. Ultrasound is directed at the heart of the baby and the frequency of the reflected beam is monitored. When the heart wall moves towards the transducer the frequency will increase, and when it moves away the frequency will fall. Thus, the heart rate can be found from the number of frequency changes per minute (Figure 32).

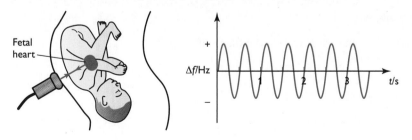

Figure 32

Knowledge check 10

The frequency of the engine noise of a racing car engine is 100 Hz when the car is travelling at 68 m s^{-1}. A spectator at the side of the track hears a sound of frequency 120 Hz as the car approaches. What engine frequency will the observer hear after the car has passed, and is moving away?

Summary

After studying this section, you should be able to:

- describe how waves are refracted in terms of a change of wave speed across a boundary
- use Snell's law to determine relative and absolute refractive indices, and to calculate the critical angle at an interface
- understand the concept of plane polarised waves, and describe how polarised light is produced and detected
- state the meaning of diffraction, and describe how diffraction is affected by the wavelength of the radiation and the size of the aperture or obstruction
- use pulse-echo techniques to calculate distances or wave speeds, and describe how the frequency changes when the source of a wave moves relative to an observer

DC electricity

Current, I = rate of flow of charge

Potential difference, V = work done in transferring 1 coulomb of charge

Resistance, R = potential difference divided by current

An **electric current** is a flow of charge, i.e. a movement of charge carriers such as electrons and ions. Direct current (DC) flows in one direction only. In order to make charges move, work must be done. A 'voltage', or more correctly a **potential difference**, between two points is needed to provide this energy. When a current flows in a circuit, it will encounter a **resistance** to the flow.

These quantities are represented by the following equations:

$$I = \frac{\Delta Q}{\Delta t} \quad \text{unit: ampere (A)}$$

$$V = \frac{W}{Q} \quad \text{unit: volt (V)}$$

$$R = \frac{V}{I} \quad \text{unit: ohm } (\Omega)$$

Worked example

(a) How much charge flows through a filament lamp in 1 minute when the current is 500 mA?

(b) Given that the electron charge is 1.6×10^{-19} C, how many electrons flow during this time?

(c) What is the potential difference across the filament if 60 J of energy is emitted by the filament?

(d) Calculate the resistance of the filament.

Answer

(a) $\Delta Q = I \Delta t = (500 \times 10^{-3}\,\text{A}) \times 60\,\text{s} = 30\,\text{C}$

(b) Number of electron $= \dfrac{\text{total charge}}{\text{charge on each electron}} = \dfrac{30\,\text{C}}{1.6 \times 10^{-19}\,\text{C}} = 1.9 \times 10^{20}$

(c) $V = \dfrac{W}{Q} = \dfrac{60\,\text{J}}{30\,\text{C}} = 2.0\,\text{V}$

(d) $R = \dfrac{V}{I} = \dfrac{2.0\,\text{V}}{0.50\,\text{A}} = 4.0\,\Omega$

Examiner tip
The expressions $V = IR$ and $I = V/R$ are useful arrangements of the definition of resistance, but they are not accepted as definitions of potential difference and current.

Series circuits

A series circuit is a circuit in which the components are connected in sequence, one after the other (Figure 33).

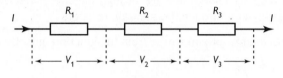

Figure 33

In all circuits, the laws of conservation of charge and conservation of energy must apply. In the diagram above, three resistors are connected in series. For charge to be conserved, the rate of flow of charge must be the same through each resistor. It follows that *in series circuits the current, I, must be the same in all components.*

For conservation of energy to hold, the total energy transferred to the resistors must equal the sum of the energies transferred in each component. Therefore, the total work done per coulomb transferred will equal the sum of the work done per coulomb transferred in each of the resistors. Since work done per coulomb transferred is the same as potential difference, we have:

$$V_{total} = V_1 + V_2 + V_3$$

Using $V = IR$ from the definition of resistance and the fact that I is the same throughout the circuit, the above equation becomes $IR_{total} = IR_1 + IR_2 + IR_3$. Cancelling out I gives the total resistance in the series circuit as:

$$R_{total} = R_1 + R_2 + R_3$$

Because the current is the same at all parts of a series circuit, an ammeter connected in series will give the same reading no matter where it is placed relative to the resistors. For the ammeter itself to have little effect on the overall resistance of the circuit, it needs to have an extremely small resistance.

Parallel circuits

In a parallel circuit, every component is connected to the same two points (Figure 34).

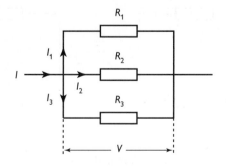

Figure 34

From conservation of charge it follows that the total charge entering a junction must be the same as that leaving the junction, so:

$$Q = Q_1 + Q_2 + Q_3$$

where Q denotes the total charge flowing through the circuit and Q_1, Q_2, Q_3 are the amounts of charge flowing through the separate resistors.

The rates of flow of charge into and away from the junction must then also be equal, giving:

$$I = I_1 + I_2 + I_3$$

The law of conservation of energy requires that the ratio W/Q be the same for all three resistors; in other words, the potential difference must be the same across all three resistors: $V_1 = V_2 = V_3 = V$. Using the definition of resistance in the rearranged form $I = V/R$, we get:

$$\frac{V}{R} = \frac{V}{R_1} + \frac{V}{R_2} + \frac{V}{R_3}$$

and hence:

$$\frac{1}{R} = \frac{1}{R_1} + \frac{1}{R_2} + \frac{1}{R_3}$$

Worked example

(1) Calculate the total resistance of each of the three circuits below.

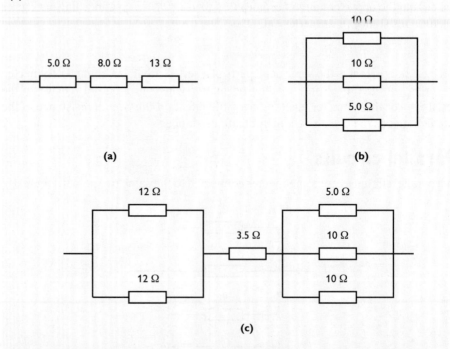

(2) A potential difference of $6.0\,V$ is applied across the ends of circuit (c) above. Calculate: (i) the total current that flows in the circuit, (ii) the potential difference across the $5.0\,\Omega$ resistor, (iii) the current flowing in the $5.0\,\Omega$ resistor.

Answer

(1) (a) Three resistors in series, so $R = 5.0\,\Omega + 8.0\,\Omega + 13\,\Omega = 26\,\Omega$

(b) Three resistors in parallel, so $\dfrac{1}{R} = \dfrac{1}{10\,\Omega} + \dfrac{1}{10\,\Omega} + \dfrac{1}{5.0\,\Omega} = 0.40\,\Omega^{-1}$ and hence $R = 2.5\,\Omega$

(c) We need to find the resistance of each of the two parallel networks and then add the three resistances in series.

For the parallel circuit on the left, $\dfrac{1}{R} = \dfrac{1}{12\,\Omega} + \dfrac{1}{12\,\Omega}$ gives $R = 6.0\,\Omega$

For the parallel circuit on the right, $\dfrac{1}{R} = \dfrac{1}{5.0\,\Omega} + \dfrac{1}{10\,\Omega}$ gives $R = 2.5\,\Omega$

The total resistance is $6.0\,\Omega + 3.5\,\Omega + 2.5\,\Omega = 12\,\Omega$

(2) (i) $I = \dfrac{V}{R} = \dfrac{6.0\,\text{V}}{12\,\Omega} = 0.50\,\text{A}$

(ii) The potential difference across the $5.0\,\Omega$ resistor is the same as the potential difference across the whole right-hand side circuit of three resistors in parallel. Now, the current flowing through the right-hand side parallel circuit is the same as the total current that flows in the entire circuit; so, using $V = IR$ for the right-hand side parallel circuit, we get:

potential difference = total current × resistance of right-hand side parallel circuit

$V = 0.50\,\text{A} \times 2.5\,\Omega = 1.25\,\text{V}$

(iii) $I = \dfrac{V}{R} = \dfrac{1.25\,\text{V}}{5.0\,\Omega} = 0.25\,\text{A}$

Voltmeters measure the potential difference across components, therefore they must be connected in parallel with the components. Some current will be diverted from the circuit by the voltmeter. For this amount of current to be negligible, the voltmeter needs to have a very large resistance.

Electrical energy and power

You have seen that in order for a current to flow, work needs to be done on the charge carriers. This is expressed in the definition of potential difference, which says that $V = W/Q$ or, more accurately, since the denominator is 'amount of charge transferred', $V = W/\Delta Q$. Rearranging and combining with the formula $I = \Delta Q / \Delta t$ gives:

$W = V\Delta Q = VI\Delta t$

When a current passes through a resistor, the work done converts electrical energy to thermal energy in the resistor. For a steady current, the **electrical energy** transferred to a circuit in time t is therefore:

$E = VIt$

Worked example

The heating element of the rear window of a car has a resistance of $2.0\,\Omega$ and is connected to a 12 V battery. How much energy does it transfer in 20 minutes?

> **Knowledge check 11**
>
> Calculate the total resistance of three $15\,\Omega$ resistors connected in parallel, in series with a pair of $18\,\Omega$ resistors that are also connected in parallel.

> **Electrical energy** is the work done by a potential difference, V, when a charge ΔQ is transferred. The energy transferred is the product $V\Delta Q$.

Answer

$$I = \frac{V}{R} = \frac{12\,V}{2.0\,\Omega} = 6.0\,A$$

$$W = VIt = 12\,V \times 6.0\,A \times (20 \times 60\,s) = 86\,kJ$$

Electrical power is the rate at which electrical energy is dissipated in a circuit. It is usually defined by the equation $P = VI$

In Unit 1, **power** was defined as the rate of doing work:

$$P = \frac{\Delta W}{\Delta t}$$

Hence, for a steady current:

$$P = \frac{VIt}{t} = VI$$

Using the definition of resistance, $R = V/I$, this can also be written as:

$$P = I^2 R \text{ or } P = \frac{V^2}{R}$$

Worked example

The power from a 10 MW wind farm is transmitted at a potential difference of 330 000 V.

(a) Calculate the current that flows when the farm is generating at its maximum capacity.

(b) The electricity is transferred to a sub-station through overhead cables of resistance 100 Ω.

Calculate the power loss in the cables.

(c) A student suggested that it would be safer and more efficient to transmit the power at 100 000 V.

Determine the power that would be lost in the cables at this voltage of transmission, and comment on the student's suggestion.

Answer

(a) $P = V \times I \Rightarrow I = \dfrac{P}{V} = \dfrac{10^7\,W}{3.3 \times 10^5\,V} = 30\,A$ (to 2 s.f.)

(b) $\Delta P = I^2 R = (30\,A)^2 \times 100\,\Omega = 90\,kW$

(c) At a transmission voltage of $10^5\,V$, the current I would be $\dfrac{10^7\,W}{10^5\,V} = 100\,A$, so the power loss would be $\Delta P = I^2 R = (100\,A)^2 \times 100 = 1.0\,MW$.

At a lower voltage the current will be larger and, since the power loss in the cables is dependent on the square of the current, much more energy will be transferred to the surroundings when the transmission voltage is lower. For this reason, the national grid transmits at as high a voltage as possible.

The dangers of high voltages are a problem, of course, and the voltage is stepped down for domestic use. However, even from a safety perspective there is little to be gained by dropping the transmission voltage from 330 000 V to 100 000 V.

> **Knowledge check 12**
>
> Calculate the energy given out by an electric kettle element of resistance 20 Ω in 5.0 minutes. The mains voltage is 230 V.

Ohm's law

Ohm's law is a statement on how the current in a metallic conductor relates to the potential difference across the ends of the conductor.

Some non-metallic components can behave in the same way as metals at a fixed temperature. Such conductors, in which the current is proportional to the voltage, are said to be **ohmic**.

The characteristics of a number of different conductors can be studied by applying a range of potential differences across them, measuring the corresponding currents and plotting the associated I–V graphs. This may be achieved by using ammeters and voltmeters, or, if the precision of the readings is not important, the data may be fed into a computer interface and the graphs displayed electronically (Figure 35).

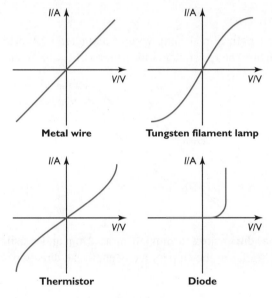

Figure 35

For ohmic conductors, I is always proportional to V so that the ratio of potential difference to current (V/I) is constant. This means that the resistance of an ohmic conductor is the same at all values of potential difference and current. In other, non-ohmic, components the resistance is variable. From the graphs in Figure 35, it can be seen that in the filament lamp, the ratio V/I (i.e. the resistance) increases as the current grows. For the thermistor, on the other hand, resistance, V/I, decreases as the current rises so the line's slope decreases. The resistance in these components is affected by the temperature, a phenomenon that will be discussed later. Diodes are designed to allow the current to flow in one direction only — they have an extremely high resistance when 'reverse biased', and become conducting only after a small forward potential difference is applied.

Resistivity

If asked whether copper wire has lower resistance than nichrome wire, it is tempting to answer yes, as copper is used for connecting leads and nichrome for heating

Examiner tip
A common error is to take one of the equations for the definition of resistance ($R = V/I$, $I = V/R$ or $V = IR$) as the statement of the law. This is not valid, because these equations hold in all conditions, whereas Ohm's law applies only to metals at constant temperature.

elements. However, if you measured the resistance of a 20 m length of copper wire of diameter 0.5 mm, the value would turn out to be greater than the resistance of a 10 cm-long nichrome rod of diameter 1 cm.

Simple measurements show that the resistance of a piece of wire is directly proportional to its length and inversely proportional to its cross-sectional area:

$$R = \frac{\rho l}{A}$$

The constant of proportionality, ρ, is dependent upon the material from which the wire is made and is known as the **resistivity** of the material.

$$\text{resistivity} = \frac{\text{resistance} \times \text{cross-sectional area}}{\text{length}}$$

$$\rho = \frac{RA}{l} \quad \text{unit: } \Omega\text{m}$$

Resistivity is a property of the material.
Resistivity varies with temperature.

Worked example

(a) Calculate the length of nichrome wire of diameter 1.2 mm that is needed to make a 5.0 Ω resistor. The resistivity of nichrome is $1.1 \times 10^{-6}\,\Omega\text{m}$.

(b)

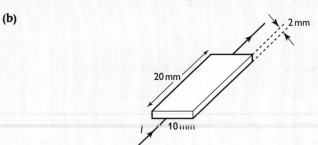

A graphite chip has dimensions 20 mm × 10 mm × 2 mm and resistivity $7 \times 10^{-5}\,\Omega\text{m}$. Calculate the resistance of the chip for a current in the direction shown.

Answer

(a) $l = \dfrac{RA}{\rho} = \dfrac{5.0\,\Omega \times \pi(0.6\times10^{-3}\,\text{m})^2}{1.1\times10^{-6}\,\Omega\text{m}} = 5.1\,\text{m}$

(b) $R = \dfrac{\rho l}{A} = \dfrac{(7\times10^{-5}\,\Omega\text{m})\times(20\times10^{-3}\,\text{m})}{(10\times10^{-3}\,\text{m})\times(2\times10^{-3}\,\text{m})} = 0.07\,\Omega$

Potential dividers

In DC circuits it is often necessary to apply a potential difference across a component that is less than the voltage of the supply. It is quite simple to use a pair of resistors connected in *series* to divide the voltage by 'sharing' it between the resistors (Figure 36).

The ratio of the voltages across the two resistors can be written as:

$$\frac{V_1}{V_2} = \frac{IR_1}{IR_2} = \frac{R_1}{R_2}$$

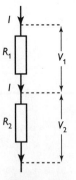

Figure 36

since I is the same in each resistor.

It is common to show a potential divider giving a required output voltage V_O from a supply of voltage V_S. In this case, let $V_O = V_2$ and $V_S = V_1 + V_2$, which leads to:

$$\frac{V_o}{V_s} = \frac{R_2}{R_1 + R_2}$$

and hence:

$$V_O = \frac{R_2}{R_1 + R_2} \times V_S$$

Examiner tip

The principle of a potential divider is that the ratio of the voltages across the resistors connected in series is equal to the ratio of their resistances.

Worked example

(a)

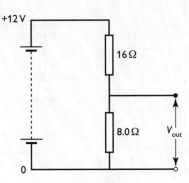

(b)

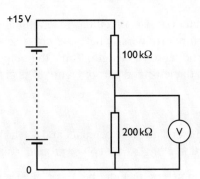

(a) Calculate the output voltage across the $8.0\,\Omega$ resistor.

(b) (i) Calculate the output voltage in the circuit.

    (ii) A voltmeter of resistance $200\,k\Omega$ is connected across the output. What reading will it give for the output voltage?

Answer

(a) $V_{out} = \dfrac{8.0\,\Omega}{16\,\Omega + 8.0\,\Omega} \times 12\,V = 4.0\,V$

(b) (i) $V_{out} = \dfrac{200\,k\Omega}{100\,k\Omega + 200\,k\Omega} \times 15\,V = 10\,V$

    (ii) The voltmeter and the $200\,k\Omega$ resistor in parallel will give a combined resistance of $100\,k\Omega$. Hence:

$$V_O = \frac{100\,k\Omega}{100\,k\Omega + 100\,k\Omega} \times 15\,V = 7.5\,V$$

Note: this is an example of how using an instrument to measure the value of a quantity can affect that measured value. In most cases, especially with digital meters, the voltmeter has such a high resistance that it takes virtually no current from the network, thus leaving the voltage almost unaffected.

Potentiometers

A potentiometer can be used to obtain a continuously variable output voltage (Figure 37).

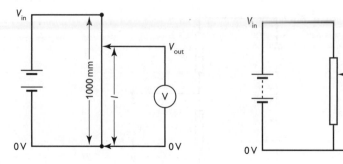

Figure 37

Examiner tip

Potentiometers are basically potential dividers in which the resistors can be continuously varied.

Knowledge check 14

A 12V battery is connected to the ends of a 1.00m length of nichrome wire. Calculate the potential difference across 40cm of the wire.

The simplest potentiometer consists of a uniform length of resistance wire stretched along a metre rule. For uniform wire (where the cross-sectional area is constant), the resistance is proportional to the length. Thus, the output voltage for the simple potentiometer shown in the left-hand part of Figure 37 is given by.

$$V_{out} = \frac{l}{1000} \times V_s$$

Most practical potentiometers are more compact and use a sliding contact to vary the length of the output section along a wire coil or around a circular conductor.

Applications of potentiometers include heat- or light-controlled switches. You have met thermistors, which have resistances that change with temperature. If a thermistor or a light-dependent resistor (LDR) is connected in series with a resistor, the output voltage (across the resistor) can be controlled by temperature or light variations. The resistor is chosen so that an alarm will be triggered at a specific output voltage. Alternatively, a voltmeter across the output can be calibrated according to temperature or luminous intensity (Figure 38).

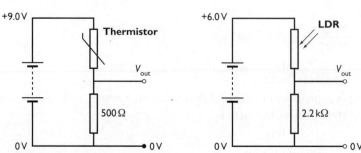

Figure 38

Edexcel AS Physics

Worked example

In the circuit shown in Figure 38, the resistance of the thermistor is 2.0 kΩ at 20°C and 400 Ω at 50°C. Calculate the output voltage at each of these temperatures.

Answer

At 20°C, $V_O = \dfrac{500\,\Omega}{2000\,\Omega + 500\,\Omega} \times 9.0\,\text{V} = 1.8\,\text{V}$

At 50°C, $V_O = \dfrac{500\,\Omega}{400\,\Omega + 500\,\Omega} \times 9.0\,\text{V} = 5.0\,\text{V}$

Electromotive force

At the beginning of this section you were given a definition of potential difference in terms of the work done, or energy converted, within a circuit or component. The idea that a source of electrical energy is needed before a current can flow was also introduced.

The term **electromotive force**, commonly written as emf (or e.m.f.) is always used in relation to devices that convert energy from other forms into electrical energy.

$\mathcal{E} = \dfrac{\Delta E}{\Delta Q}$ unit: volt (V)

Examples of electrical sources include:

- dynamo — converts kinetic energy into electrical
- battery — converts chemical energy into electrical
- solar cell — converts radiant (light) energy into electrical

> The **emf** of a generator is the energy converted into electrical energy per coulomb of charge produced.

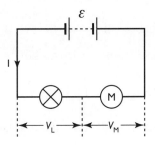

Figure 39

By the law of conservation of energy, the chemical energy converted per coulomb of charge in the battery in Figure 39 must be equal to that transformed into heat and light in the lamp and into kinetic energy in the motor.

emf of battery = potential difference across lamp + potential difference across motor

> **Examiner tip**
> Electromotive force is only used for devices that generate electricity (e.g. batteries and dynamos). In all other cases the 'voltage' across a component is called the potential difference.

Internal resistance

Unfortunately, not all of the chemical energy in the battery shown in Figure 39 will be transferred to the lamp and the motor. Some of the energy is used to 'push' the charges through the cells of the battery. These cells have a resistance between the

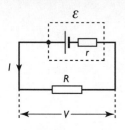

Figure 40

Examiner tip

The equation $V = \varepsilon - Ir$ is very useful in the practical determination of the internal resistance of a cell. If the potential difference across the cell is measured for a range of currents, a graph of V against I results in a straight line of gradient $-r$ with the intercept on the V axis equalling the emf of the cell.

Examiner tip

Check that using $V = \varepsilon - Ir$ gives the same result.

Knowledge check 15

Three identical cells of emf 2.0 V and internal resistance 0.50 Ω are connected such that two are in parallel, and these are then connected in series with the third cell. Calculate (a) the total emf and (b) the internal resistance of the combination.

electrodes. This is called the **internal resistance** of the cell and is denoted by r (Figure 40).

Applying the law of conservation of energy as before, we have:

emf of cell = potential difference across resistor + work done per coulomb within the cell

$$\varepsilon = V + \Delta V$$

It follows that the **terminal potential difference** V (= IR) will be less than the emf of the cell, and the difference ΔV, i.e. the 'lost volts', will equal Ir:

$$\varepsilon = IR + Ir$$

In order to calculate the terminal potential difference, it is useful to remember two rearrangements of the above equations:

- $V = \varepsilon - Ir$
- $I = \dfrac{\varepsilon}{R + r}$

Worked example

In daylight, a solar cell has an emf of 2.4 V and an internal resistance of 20 Ω. Calculate the terminal potential difference when the cell is connected to a resistor of resistance:

(a) 20 Ω

(b) 180 Ω

(c) 980 Ω

Answer

(a) $I = \dfrac{2.4\,V}{20\,\Omega + 20\,\Omega} = 0.060\,A \Rightarrow V = IR = 0.060\,A \times 20\,\Omega = 1.2\,V$

(b) $I = \dfrac{2.4\,V}{180\,\Omega + 20\,\Omega} = 1.2 \times 10^{-2}\,A \Rightarrow V = 1.2 \times 10^{-2}\,A \times 180\,\Omega = 2.2\,V$

(c) $I = \dfrac{2.4\,V}{980\,\Omega + 20\,\Omega} = 2.4 \times 10^{-3}\,A \Rightarrow V = 2.4 \times 10^{-3}\,A \times 980\,\Omega = 2.35\,V$

The above worked example shows that the terminal potential difference approaches the emf for external resistances that are much greater than the internal resistance. For extremely large external resistances, such as those of digital voltmeters, the current, and therefore the value of Ir, is virtually zero; thus the voltage across the terminals equals the emf of the cell.

For a battery of cells connected together in series, the total emf will equal the sum of the individual emfs, but the internal resistance of the battery will also be the sum of the component resistances.

For a number of identical cells connected in parallel, the emf will be the same as that of each cell, but the internal resistance will be reduced.

Factors affecting current flow in conductors

It was stated at the start of this section that an electric current is a flow of charge pushed around a circuit by a potential difference, and that work is done against the resistance of the conducting medium. You will need to have a more detailed knowledge of the mechanics of this process on the atomic level.

For *metallic* conductors, the charge is carried by **delocalised** or **free electrons**. These electrons are not fixed to specific atoms and are free to move around the atomic lattice. At room temperature, they have thermal energy and move at high speeds in a random manner, similar to gas particles in the atmosphere. Typically, there is about one free electron for every atom. When a potential difference is applied to a metallic conductor, the free electrons are forced to 'drift' through the lattice. But the motion of these electrons is impeded by their collisions with the lattice at positions where there are impurities or imperfections in the structure. This is how the resistive effect arises and, as the electrons transfer their energy to the lattice, the temperature of the metal will rise.

The magnitude of the current is given by the expression:

$I = nqvA$

where:

- n is the number of charge carriers per cubic metre (the carrier concentration)
- q is the charge on each charge carrier
- v is the drift velocity of the charge carriers
- A is the cross-sectional area of the conductor

Worked example

A tungsten filament lamp carries a current of 240 mA. The diameter of the filament is 0.024 mm and that of the copper connecting leads is 0.80 mm.

Calculate the drift velocity of the electrons in:

(a) the filament

(b) the connecting leads

You may take the number of free electrons per unit volume to be 8.0×10^{28} m^{-3} for copper and 4.0×10^{28} m^{-3} for tungsten.

Answer

(a) $v = \dfrac{I}{nqA}$

$$= \frac{0.240\,\text{A}}{(4.0 \times 10^{28}\,\text{m}^{-3}) \times (1.6 \times 10^{-19}\,\text{C}) \times \pi(0.012 \times 10^{-3}\,\text{m})^2} = 0.083\,\text{m s}^{-1}$$

(b) $v = \dfrac{0.240\,\text{A}}{(8.0 \times 10^{28}\,\text{m}^{-3}) \times (1.6 \times 10^{-19}\,\text{C}) \times \pi(0.40 \times 10^{-3}\,\text{m})^2} = 3.7 \times 10^{-5}\,\text{m s}^{-1}$

The above calculation shows how slowly the electrons drift through the lattice. Although they can have random speeds averaging several hundred metres per second, it would take the electrons about 30 seconds to drift 1 millimetre in the copper leads.

Variations in resistivity

The equation $I = nqvA$ can be used to explain the variations in resistivity of different materials at different temperatures. For a given cross-section, the current depends on the carrier concentration and the drift velocity (in most cases q equals the electron charge 1.6×10^{-19} C).

The equation tells us that the higher the carrier concentration, the better the conductivity. For example, the value of n for copper is about 8.0×10^{28} m^{-3} while that of the high-resistivity alloy nichrome is about 5.0×10^{26} m^{-3}.

Semiconducting materials such as silicon have much lower carrier concentrations — of the order of 10^{23} m^{-3} — and their resistivities are correspondingly higher than those of metals.

Resistivity also depends on temperature. The I–V curve for a tungsten filament lamp in Figure 35 illustrates the effect of temperature on the resistance of the tungsten filament. At higher voltages, when the filament is hot, the graph shows that the ratio of V to I rises and so the resistance increases.

Metals are said to have a *positive* **temperature coefficient of resistance**. At higher temperatures, owing to increased thermal vibration of the lattice, there will be greater interaction between the free electrons and the lattice. This results in a reduced drift velocity and, consequently, a reduction in the current.

In contrast to metals, semiconductors have a *negative* temperature coefficient of resistance. Thermistors are better conductors at higher temperatures (NTC). You may be familiar with the experiment demonstrating that a glass rod is insulating when cold but becomes conducting upon being strongly heated with a Bunsen flame. Although, like in metallic conductors, the increased lattice vibration will reduce the drift velocity v, the carrier concentration n increases markedly in thermistors as the temperature rises. In some cases, an overheated semiconductor will generate even more heat as the current rises, which further reduces the resistance, leading to an avalanche effect known as 'thermal runaway' and the eventual destruction of the component through melting.

Examiner tip

When the value of the resistivity of a metal is quoted, the temperature must also be given. Many reference books give the values at 20°C.

Knowledge check 16

Explain why the drift velocity of the charge carriers in a semiconductor is greater than that in a metallic conductor of the same dimensions and carrying the same current.

Knowledge check 17

State the difference between positive and negative coefficients of resistivity.

Summary

After studying this section, you should be able to:

- define current, potential difference and resistance, and use the relationships between them to determine values of these quantities in series and parallel circuits
- use the equation $R = \dfrac{\rho l}{A}$ and explain the difference between resistance and resistivity
- state Ohm's law and distinguish between ohmic and non-ohmic conductors

- understand the concepts of electromotive force and internal resistance of a source of electrical energy, and describe how the internal resistance of a cell or battery is determined
- use the equation $I = nqvA$ to find the current or drift velocity in a conductor, and explain positive and negative coefficients of resistivity

Nature of light

This section concentrates mainly on the photon nature of light and the development of basic quantum concepts. You will not be examined on the historical evolution of modern-day theories but, if you have the time, a little background reading can be fascinating and may contribute to a fuller understanding of the nature of light. A brief synopsis of some of the significant stages is outlined below.

In the seventeenth century, Sir Isaac Newton and Christiaan Huygens proposed conflicting theories on the nature of light. Newton's corpuscular model was based on a study of the mechanics of particles, while Huygens suggested that light behaves as a wave. Around 1800, Thomas Young performed his famous double-slit experiment, which demonstrated the formation of interference patterns — a clear indication of the wave nature of light. When measurements of the speed of light in water were made later in the nineteenth century, the value turned out to be less than the speed of light in air, providing further evidence that contradicted Newton's theory.

Although James Clerk Maxwell's equations, published in the late nineteenth century, linked the variations of electric and magnetic fields with the speed of light, there were still some phenomena that could not be explained using the classical wave model.

At the start of the twentieth century, Max Planck suggested that the reduction in peak wavelength of emissions of electromagnetic radiation at higher temperatures (e.g. the colour of hot objects glowing from red to orange to white as the temperature is increased) was consistent with the light being emitted in 'bundles' or 'quanta' of energy, with each package having an energy that is dependent on the frequency of the wave.

After Albert Einstein used Planck's ideas to explain the photoelectric effect and Niels Bohr adapted Rutherford's atomic model to predict spectral emissions, the photon nature of light became universally accepted.

The final link was made by Louis de Broglie, who showed that photons exhibit both wave and particle properties, thus indicating that Newton's theory was valid after all.

Examiner tip

It is important to understand that light, and many other electromagnetic radiations, consists of photons, and that the properties of the radiation depend on the energy of each photon.

Planck's equation

A photon is a 'quantum' of wave energy. You can think of it as a short burst of electromagnetic waves emitted in a time of about 1 ns. A source of light will give out vast numbers of photons every second.

The energy carried by each photon is given by:

$$E = hf$$

where f is the frequency of the radiation and h is **Planck's constant**, with a value of 6.63×10^{-34} J s.

Worked example

Calculate:

(a) the energy in one photon of blue light of frequency $6.92 \times 10^{14}\,Hz$

(b) the wavelength of a photon with energy $3.44 \times 10^{-19}\,J$

Answer

(a) $E = hf = (6.63 \times 10^{-34}\,J\,s) \times (6.92 \times 10^{14}\,s^{-1}) = 4.6 \times 10^{-19}\,J$

(b) $E = hf = \dfrac{hc}{\lambda}$

Hence $\lambda = \dfrac{hc}{E} = \dfrac{(6.63 \times 10^{-34}\,J\,s) \times (3.00 \times 10^{8}\,m\,s^{-1})}{3.44 \times 10^{-19}\,J} = 5.78 \times 10^{-7}\,m$

> An **electronvolt** is the work done on an electron in moving it through a potential difference of 1 volt.

From the above calculation, we found that the energy in a photon of blue light is about $4.6 \times 10^{-19}\,J$. This is a very small amount of energy, and it is often more convenient to use a non-SI unit, the **electronvolt** (eV).

Recall the definition of potential difference, from which we have $W = VQ$. So:

$$1\,eV = 1\,V \times (1.6 \times 10^{-19}\,C) = 1.6 \times 10^{-19}\,J$$

> **Knowledge check 18**
>
> Calculate the photon energy of (a) radiation of frequency $7.2 \times 10^{15}\,Hz$ and (b) radiation of wavelength $6.3 \times 10^{-11}\,m$.

In many instances, photons are emitted when charges are accelerated by a potential difference, in which case the electronvolt is a particularly convenient energy unit. For example, in X-ray tubes, electrons are accelerated towards the anode by voltages of around 100 kV (for diagnosis) or about 10 MV (for therapy). In such cases the energy of the electrons, and of the resultant X-rays, is typically given in keV or MeV.

> **Examiner tip**
>
> When using Planck's equation $E = hf$ for calculations, the electronvolt energy must always be converted into joules (by multiplying by 1.6×10^{-19}).

Worked example

Calculate the wavelength of a 120 keV X-ray.

Answer

$$E = 120 \times 10^{3}\,eV \times (1.6 \times 10^{-19}\,J\,eV^{-1}) = 1.92 \times 10^{-14}\,J$$

$$\lambda = \frac{hc}{E} = \frac{(6.63 \times 10^{-34}\,J\,s) \times (3.00 \times 10^{8}\,m\,s^{-1})}{1.92 \times 10^{-14}\,J} = 1.0 \times 10^{-11}\,m$$

Photoelectric effect

The photoelectric effect refers to the emission of electrons from a material (usually a metal or a metallic oxide) when light is shone onto its surface. Classical wave theory was unable to explain why certain wavelengths of light were able to release electrons, even at very low intensity, while other wavelengths failed to eject any electrons, regardless of the intensity of the incident radiation.

> The **work function** of a material is the minimum amount of energy a photon requires to release an electron from the surface of the material.

Einstein used Planck's equation and the law of conservation of energy to provide a simple explanation. The electrons require a minimum amount of energy, known as the **work function** ϕ, to release them from the metal.

If a photon of incident light transfers its energy to an electron, the electron will be dislodged only if the photon energy is equal to or greater than the work function. If the photon energy is greater than the work function, the excess energy will be transferred as kinetic energy of the emitted electron. This can be expressed as follows:

energy of photon = work done to eject electron + kinetic energy of emitted electron

$$hf = \phi + \tfrac{1}{2}mv^2_{max}$$

The work function varies for different elements. Only 1.9 eV is needed to release an electron of caesium, whereas 4.3 eV is needed for zinc and about 5.0 eV for copper.

The minimum frequency of incident radiation needed to release a photoelectron is called the **threshold frequency**, denoted by f_0. In this case, all of the photon energy is used to liberate the electron, with none left over to transfer as kinetic energy. In Einstein's equation, the kinetic energy term will be zero, so the work function ϕ equals hf_0.

Worked example

When visible light is shone onto a polished zinc surface, no photoelectric emission is observed. However, electrons are emitted from the zinc when ultraviolet light of wavelength 200 nm is used. Explain why this occurs, given that the work function of zinc is 4.3 eV.

Answer

$$\phi = hf_0 = 4.3\,\text{eV} = 4.3\,\text{eV} \times (1.6 \times 10^{-19}\,\text{J eV}^{-1}) = 6.88 \times 10^{-19}\,\text{J}$$
$$f_0 = \frac{\phi}{h} = \frac{6.88 \times 10^{-19}\,\text{J}}{6.63 \times 10^{-34}\,\text{J s}} = 1.0 \times 10^{15}\,\text{Hz}$$

The shortest waves in the visible spectrum are at the blue end, with a wavelength of about 400 nm. From $c = f\lambda$ we find that the corresponding frequency is:
$$f = \frac{3.0 \times 10^8\,\text{m s}^{-1}}{400 \times 10^{-9}\,\text{m}} = 7.5 \times 10^{14}\,\text{Hz}$$

This is less than the threshold frequency for zinc, so no photoelectric emission occurs when visible light is shone onto the zinc surface.

For ultraviolet light, $f = \dfrac{3.0 \times 10^8\,\text{m s}^{-1}}{200 \times 10^{-9}\,\text{m}} = 1.5 \times 10^{15}\,\text{Hz}$.

This frequency is greater than f_0 and so the ultraviolet photons have sufficient energy to remove the electrons.

The kinetic energy of photoelectrons can be found by measuring their **stopping potential** (Figure 41).

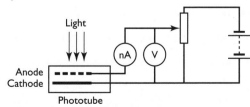

Figure 41

Examiner tip

The equation $hf = \phi + \tfrac{1}{2}mv^2_{max}$ represents the case where all of the residual photon energy is converted into kinetic energy of the electron. In practice, some electrons may require extra energy to reach the surface, so their kinetic energy will be less than the maximum value given in the equation.

Examiner tip

Most common metals have work functions greater than the energy of photons of visible light, and so need ultraviolet radiation for photoelectric emission.

Knowledge check 19

Calculate the work function of a metal if the maximum kinetic energy of the photoelectrons emitted by a photon of energy 9.3×10^{-19} J is 1.3×10^{-19} J. Give your answer in eV.

A phototube is an evacuated glass envelope containing an anode and a cathode. The cathode is coated with caesium so that when light is incident upon it, photoelectrons will be emitted. A nano-ammeter connected across the electrodes will detect a small current as the electrons move from cathode to anode. If a reverse potential difference is applied (i.e. if the cathode is connected to the positive terminal of a battery), then the electrons will be slowed down. As the potential difference is increased, more work is done on the electrons; when the work done becomes equal to the maximum kinetic energy of the electrons, they will stop moving and the current on the ammeter will read zero. The minimum voltage required to stop all of the electrons is called the stopping potential, denoted by V_s. From the relation $W = QV$, the minimum work done on an electron to stop it moving is eV_s where e is the charge on a single electron, 1.6×10^{-19} C. Thus we have the equation:

$$eV_s = \frac{1}{2}mv^2_{\text{max}}$$

Einstein's equation can now be written as:

$$hf = hf_0 + eV_s$$

Examiner tip

If Einstein's equation is rearranged as
$eV_s = hf - hf_0$
you can see that a graph of the maximum KE (eV_s) against the frequency of the incident light (f) will give a straight line of gradient h, intersecting the energy axis at $-hf_0$. When $eV_s = 0$, $f = f_0$, and so the line will cross the frequency axis at the threshold frequency.

Worked example

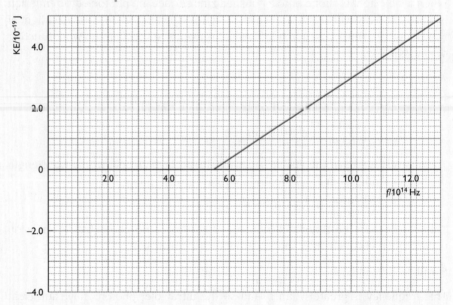

The above graph shows how the maximum kinetic energy, KE, of photoelectrons emitted from the surface of sodium metal varies with the frequency, f, of the incident electromagnetic radiation.

(a) Use the graph to find a value for the Planck constant.

(b) Use the graph to find the work function, ϕ, of sodium metal.

(c) Calculate the stopping potential when the frequency of the incident radiation is 9.0×10^{14} Hz.

Answer

(a) From $\frac{1}{2}mv^2_{max} = hf - \phi$, we can see that h is the gradient of the graph, so

$$h = \frac{(4.9 - 0.0) \times 10^{-19}\,J}{(13.0 - 5.5) \times 10^{14}\,s^{-1}} = 6.5 \times 10^{-34}\,Js$$

(b) When $\frac{1}{2}mv^2_{max} = 0$

$hf_0 = \phi$, where f_0 = threshold frequency = $5.5 \times 10^{14}\,Hz$

$\phi = 6.5 \times 10^{-34}\,Js \times 5.5 \times 10^{14}\,s^{-1} = 3.6 \times 10^{-19}\,J\ (= 2.2\,eV)$

(c) $\frac{1}{2}mv^2_{max} = hf - \phi = (6.5 \times 10^{-34}\,J\,s \times 9.0 \times 10^{14}\,s^{-1}) - 3.6 \times 10^{-19}\,J = (5.9 - 3.6) \times 10^{-19}\,J$

$$= 2.3 \times 10^{-19}\,eV$$

$$V_s = \frac{2.3 \times 10^{-19}\,J}{1.6 \times 10^{-19}\,J} = 1.4\,V$$

Atomic spectra

When gases or vaporised elements are heated to high temperatures, or have a very large potential difference applied across them, they emit electromagnetic radiation consisting of a number of characteristic wavelengths. This group of frequencies is often shown as as lines on a diffraction image, and is called the **emission spectrum** for that element.

Rutherford's model of an atom, consisting of a positive nucleus with electrons orbiting like the planets around the Sun, was a major step forward in explaining atomic behaviour. However, classical physics predicted that the electrons ought to collapse inwards because a charged particle in a circular orbit accelerates towards the centre of the orbit, and accelerating charges give off electromagnetic radiation, thus losing energy. By applying quantum theory to a model of the hydrogen atom, Niels Bohr showed that an electron could only exist in certain discrete or **quantised** orbits. The lowest-energy state — the most stable orbit — is called the **ground state**. If energy is given to an electron, it can exist in a number of **excited** states, also called **permitted orbitals**.

The permitted orbitals are generally represented on a chart of **energy levels** labelled $n = 1, 2, 3$ and so on. The ground state corresponds to $n = 1$ and the energy difference between successive levels gets smaller as n increases. Figure 42 shows the energy levels of a hydrogen atom.

If sufficient energy is given to an electron, it can be completely removed from the atom, which is now **ionised**. For the hydrogen atom, 13.6 eV is needed to raise the electron from its ground state to the ionisation level (labelled $n = \infty$). By convention, an electron that has *just* been removed from the atom is assigned 'zero' energy, so energy levels below ionisation take negative values that indicate how much energy is needed to ionise the atom from each level.

An **emission spectrum** consists of a number of characteristic frequencies emitted by the atoms of an element when they are at high temperatures or in a strong electric field.

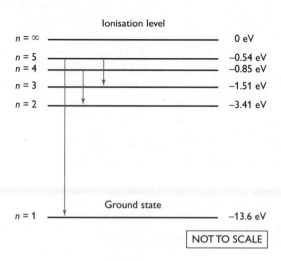

Figure 42

When an electron drops from a higher energy level to a lower level, it releases the energy difference in the form of one **quantum** of radiation, hf. Thus, when an electron falls from energy level E_1 to energy level E_2:

$$E_1 - E_2 = hf$$

The arrows in the above diagram illustrate some possible transitions between energy levels. Electrons in higher energy levels can fall to any lower level. Particular elements have characteristic differences between their atomic energy levels, therefore the emitted photons have a set of characteristic frequencies which can be used to identify the element via spectral analysis. For hydrogen, the energy emitted by the electron falling to level $n = 2$ will produce photons in the visible region, while more energetic, higher-frequency photons in the ultraviolet region will be released when the electron falls directly to the ground state.

Worked example

Use the above diagram of energy levels for the hydrogen atom to answer the following questions.

(a) Calculate the ionisation energy in joules for an electron in the −13.6 eV level.

(b) What is the wavelength of the light emitted when an electron falls from the −1.51 eV level to the −3.41 eV level? Suggest what colour the light would be.

(c) Between which energy levels must an electron fall to emit blue light having a wavelength of 434 nm?

(d) Without doing any calculations, explain why the radiation emitted when an electron falls to its lowest energy level cannot be seen.

Answer

(a) Ionisation energy = 13.6 eV = 13.6 × (1.6 × 10⁻¹⁹) J = 2.2 × 10⁻¹⁸ J

Examiner tip

Energy levels are often given in electronvolts and need to be converted to joules (multiplied by 1.6×10^{-19} J eV⁻¹) in order to calculate the frequency of the emission using Planck's equation.

(b) $hf = E_1 - E_2 = (-1.51)\,\text{eV} - (-3.41)\,\text{eV} = 1.90\,\text{eV} = 1.90\,\text{eV} \times (1.6 \times 10^{-19}\,\text{J eV}^{-1})$
$= 3.04 \times 10^{-19}\,\text{J}$

$$f = \frac{3.04 \times 10^{-19}\,\text{J}}{6.63 \times 10^{-34}\,\text{J s}} = 4.59 \times 10^{14}\,\text{Hz}$$

$$\lambda = \frac{c}{f} = \frac{3.00 \times 10^8\,\text{m s}^{-1}}{4.59 \times 10^{14}\,\text{s}^{-1}} = 6.54 \times 10^{-7}\,\text{m} = 654\,\text{nm}$$

This falls in the range of visible red light.

(c) $f = \dfrac{c}{\lambda} = \dfrac{3.00 \times 10^8}{434 \times 10^{-9}} = 6.91 \times 10^{14}\,\text{Hz}$

$E_1 - E_2 = hf = (6.63 \times 10^{-34}\,\text{J s}) \times (6.91 \times 10^{14}\,\text{s}^{-1}) = 4.58 \times 10^{-19}\,\text{J}$

$$= \frac{4.58 \times 10^{-19}\,\text{J}}{1.60 \times 10^{-19}\,\text{J eV}^{-1}} = 2.86\,\text{eV}$$

This can be achieved by the electron falling from the $-0.54\,\text{eV}$ level to the $-3.41\,\text{eV}$ level.

(d) The energy released when an electron falls to the lowest level would be much greater than the value of $2.86\,\text{eV}$ calculated above, which corresponds to blue light. The radiation emitted must therefore have a much higher frequency and thus lie beyond the visible region of the electromagnetic spectrum.

Knowledge check 20

Explain the terms (a) permitted orbital, (b) quantum jump, (c) ionisation energy.

Intensity of light

The intensity of light is the brightness or strength of the radiation that is received by an observer. The **intensity** of a star is not a measure of how much radiation per unit time is emitted by the star, but rather of how much radiation per unit time is received on Earth.

Intensity is often referred to as **radiation flux**.

$$\text{Intensity} = \frac{\text{power received}}{\text{area}}$$

$I = \dfrac{P}{A}$ unit: W m^{-2}

The **intensity** (radiant flux) is the power of the radiation falling perpendicularly onto one square metre of a surface.

Intensity depends not only on the power of the source, but also on the amount by which the beam spreads between source and observer. For example, the Sun emits radiation at a power of $4 \times 10^{26}\,\text{W}$, but this radiation is spread out in all directions and the intensity that actually reaches the Earth is just over $1\,\text{kW m}^{-2}$. By comparison, a $5\,\text{mW}$ laser emitting a beam of diameter $1.5\,\text{mm}$ can form an image of intensity around $3\,\text{kW m}^{-2}$.

A simple formula for how intensity depends on distance from the source can be derived as follows. Consider a point source that is emitting radiation at power P uniformly in all directions. The points at a fixed distance, d, from the source form a spherical shell (i.e. surface of a sphere) of radius d and surface area $A = 4\pi d^2$.

Therefore, at these points the intensity will be $I = \dfrac{P}{4\pi d^2}$. Note that this relationship between I and d is an inverse square law.

Examiner tip

The inverse square law means that if the distance of the source from the receiver is doubled, the intensity will decrease four times.

Examiner tip

A common misunderstanding is the idea that the intensity of radiation falling onto the Earth in winter is less than in the summer because the Earth in one hemisphere is further away from the Sun. The change in distance due to the inclination of the Earth is negligible compared with the distance from the Sun. It is the angle at which the radiation falls onto the surface that reduces the intensity.

Knowledge check 21

The intensity of light from a lamp falling onto a surface 0.50m away is 8.0Wm^{-2}. What will be the intensity of the light falling onto the surface if the lamp is moved to 2.00m from the surface?

Worked example

A panel of photovoltaic cells has an area of 3.2 m^2. When the Sun's rays fall normally on the panel (i.e. at 90° to the surface), the solar radiation flux is 400 W m^{-2}.

(a) Calculate how much energy falls on the panel every second:

    (i) when the Sun's rays are normal to the surface

    (ii) when the Sun's rays fall at an angle of 30° to the surface

(b) If the panel has an emf of 24 V and a maximum output current of 15 A, calculate the efficiency of the panel.

Answer

(a) (i) Power = radiation flux × area = 400 W m^{-2} × 3.2 m^2 = 1300 J s^{-1} (to 2 s.f.)

    (ii) The component of the solar flux that falls normally on the panel is 400 W m^{-2} × sin 30° = 200 W m^{-2}

    Therefore, power = 200 W m^{-2} × 3.2 m^2 = 640 J s^{-1}

(b) Electrical power output = $V \times I$ = 24 V × 15 A = 360 W

$$\text{Efficiency} = \frac{\text{useful power output}}{\text{total power input}} \times 100\% = \frac{360\,\text{W}}{1300\,\text{W}} \times 100\% = 28\%$$

Developments in photocell technology have led to solar panels with efficiencies approaching 40%, making the use of solar power as a renewable energy source more viable. It has been proposed that an array of reflecting panels in the Sahara desert could provide a large percentage of Europe's energy needs in the future.

Wave–particle duality

The wave nature of an electron was mentioned in the section on diffraction of waves. Photons of light are deviated by huge gravitational forces as they pass close to black holes, and 'radiation pressure' can be measured for a stream of photons. These observations suggest that certain 'particles', such as electrons, have wave properties, while 'waves', such as photons of light, can behave like particles.

Louis de Broglie introduced a relationship between wave and particle behaviour, suggesting that particles with momentum p would have a corresponding wavelength λ given by:

$$\lambda = \frac{h}{p}$$

Thus it would appear that Newton's corpuscular theory of light had some merit after all.

After studying this section, you should be able to:

- understand that light consists of photons, and that the energy of a photon is given by the expression $E = hf$
- recollect that photon energy is often shown to have the units electronvolts (eV), and that 1 eV is equivalent to 1.6×10^{-19} J
- use Einstein's equation to explain the photoelectric effect and use it to determine values of work function, threshold frequency, maximum electron energy and stopping potential
- explain how emission spectra are produced and calculate the frequencies of photons emitted when an electron falls from an excited state to a lower energy permitted orbital
- understand the meaning of radiation flux (intensity)
- be aware that photons have both particular and wave properties and that these are linked in de Broglie equation

Summary

Questions & Answers

The unit test

Unit Test 2 is a written paper of duration 1 hour 20 minutes. It carries a total of 80 marks. There are ten objective (multiple-choice) questions, each worth a single mark, and a further ten or eleven short and long questions worth between 3 and 12 marks each. The test is designed to cover all the elements in the specification for the unit, and all of the questions must be attempted; it is therefore very important to revise all topics prior to the examination.

The test will incorporate the assessment of 'How Science Works' — an exploration of how scientific knowledge is developed, validated and communicated by the scientific community (AO3). It will also examine the objectives AO1 (knowledge and understanding) and AO2 (application of knowledge and understanding) in equal measure, with each objective accounting for 30–40% of the total marks.

A formulae sheet is provided with the test. Copies can be downloaded from the Edexcel website or found at the end of past papers.

Command terms

Examiners use certain words that require you to respond in a particular way. You must be able to distinguish between these terms and understand exactly what each requires you to do. Some frequently used commands are shown below.

- **State** — the answer should be a brief sentence giving the essential facts; no explanation is required (nor should you give one).
- **Define** — you can use a word equation; if you use symbols, it is essential to state clearly what each symbol represents.
- **List** — simply give a series of words or terms; there is no need to write sentences.
- **Outline** — a logical series of bullet points or phrases will suffice.
- **Describe** — for an experiment, a diagram is essential; then state the main points concisely (bullet points can be used).
- **Draw** — diagrams should be drawn in section, neatly and fully labelled with all measurements clearly shown; but don't waste time — remember that this is not an art exam.
- **Sketch** — usually a graph is called for, but graph paper is not necessary (although a grid is sometimes provided); axes must be labelled and include a scale if numerical data are given; the origin should be shown if appropriate, and the general shape of the expected line or curve should be drawn clearly.
- **Explain** — use correct physics terminology and principles; the amount of detail in your answer should reflect the number of marks available.
- **Show that** — usually a value is provided (to enable you to proceed with the next part of the question) and you have to demonstrate how this value can be obtained;

you should show all your working and state your result to more significant figures than the given value contains (to prove that you have actually done the calculations).

- **Calculate** — show all your working and include units at every stage; the number of significant figures in your answer should reflect the given data, but you should keep each stage with more significant figures in your calculator to prevent excessive rounding.
- **Determine** — you will probably have to extract some data, often from a graph, in order to perform a calculation.
- **Estimate** — this means doing a calculation in which you have to make a sensible assumption, possibly about the value of one of the quantities; think — does your assumption lead to a reasonable answer?
- **Suggest** — there is often no single correct answer; credit is given for sensible reasoning based on correct physics.
- **Discuss** — you need to sustain an argument, giving evidence for and against, based on your knowledge of physics and possibly using appropriate data to justify your answer.

You should pay particular attention to diagrams, graph-sketching and calculations. Candidates often lose marks by failing to label diagrams properly, by not giving essential numerical data on sketch graphs and by not showing all the working or by omitting units in calculations.

About this section

The following two tests are made up of questions similar in style and content to those appearing in a typical Unit 2 examination. You might like to work through a entire paper in the allotted time and then check your answers; alternatively, you could separately attempt the multiple-choice section and selected longer questions to fit your revision plan. Use the fact that there are 80 minutes available for the 80 marks on the test to help you estimate how long you ought to spend on a particular question — you should be looking at about 10 minutes for the multiple-choice section and approximately a minute a mark on the other questions.

Although these sample papers resemble actual examination scripts in most respects, be aware that during the examination you will be writing your answers directly onto the paper, which is not possible for the tests in this book. It may be that you will need to copy diagrams and graphs that you would normally just write or draw onto in the real examination. If you are attempting one of these papers as a timed test, allow yourself an extra few minutes to account for this.

The answers should not be treated as model solutions because they represent the bare minimum necessary to gain the marks. In some instances, the difference between an A-grade response and a C-grade response is suggested. This is not possible, however, for the multiple-choice section and many of the shorter questions that do not require extended writing.

For question parts worth multiple marks, ticks (✓) are included in the answers to indicate where the examiner has awarded a mark. Note that half marks are not given.

Examiner's comments

Examiner comments on some questions are preceded by the icon ⓔ. They offer tips on what you need to do in order to gain full marks. Some student responses are followed by examiner's comments, indicated by the icon ⓔ, which highlight where credit is due or could be missed.

Test Paper 1

Questions 1–10

For questions 1–10, select one answer from A to D.

Questions 1 and 2 relate to the following displacement–distance graph of a progressive wave.

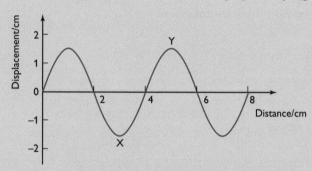

(1) The amplitude of the wave is:

A   1.5 cm

B   2.0 cm

C   3.0 cm

D   4.0 cm   (1 mark)

(2) The phase difference between points X and Y on the wave is:

A $\frac{\pi}{4}$ radians

B $\frac{\pi}{2}$ radians

C π radians

D 2π radians   (1 mark)

(3) Power can be given by the expression:

A $\dfrac{QI}{t}$

B $\dfrac{QV}{t}$

C $\dfrac{RI}{t}$

D $\dfrac{VI}{t}$ (1 mark)

(4) The resistance of 5.00 m of wire of diameter 1.0 mm and resistivity $1.1 \times 10^{-6}\,\Omega\,$m is:

A 1.8 Ω

B 5.5 Ω

C 7.0 Ω

D 22 Ω (1 mark)

(5) Which of the following waves cannot be polarised?

A light

B microwaves

C sound

D X-rays (1 mark)

(6) Coherent sources must always have the same

A amplitude

B frequency

C intensity

D phase (1 mark)

In questions 7 and 8, select the graph that best describes how the given property varies with the temperature θ.

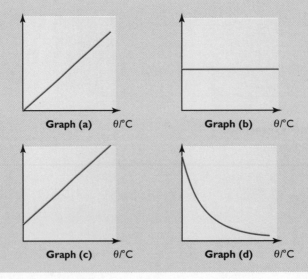

(7) The number of charge carriers per unit volume in a metal wire:

   A   graph (a)

   B   graph (b)

   C   graph (c)

   D   graph (d)   (1 mark)

(8) The resistance of an NTC thermistor:

   A   graph (a)

   B   graph (b)

   C   graph (c)

   D   graph (d)   (1 mark)

When light of wavelength 599 nm is shone onto a caesium surface, photoelectrons are emitted. The work function of caesium is 1.9 eV. Use these data to answer questions 9 and 10.

(9) The threshold frequency of caesium is:

   A 1.3×10^{-33} Hz

   B 3.0×10^{-19} Hz

   C 4.6×10^{14} Hz

   D 2.9×10^{33} Hz   (1 mark)

(10) The maximum kinetic energy of the photoelectrons is:

   A   0.18 eV

   B   1.9 eV

   C   2.1 eV

   D   4.0 eV   (1 mark)

Total: 10 marks

Answers

   (1) A

   ⓔ The amplitude is measured from the midpoint to the peak. 1.5 cm

(2) C

ⓔ X and Y are half a cycle out of phase.
Phase difference = π radians

(3) B

ⓔ Use $P = IV$ and $I = Q/t$, or $P = \dfrac{W}{t}$ and $W = QV$

(4) C

ⓔ $R = \dfrac{\rho l}{A} = \dfrac{(1.1 \times 10^{-6}\,\Omega\text{m}) \times 5.00\text{ m}}{\pi \times (0.5 \times 10^{-3}\text{m})^2} = 7.0\,\Omega$

A common error is to use the diameter rather than the radius in finding the area.

(5) C

ⓔ Sound waves are longitudinal waves, and so cannot be polarised.

(6) B

ⓔ The frequency must be the same for coherent sources; usually coherent sources are also in phase, but this is not a requirement — the phases may be different provided that there is a constant relationship between them, so D is not valid.

(7) D

ⓔ The carrier concentration in a metal remains constant at all temperatures.

(8) D

ⓔ A thermistor is a semiconducting device and so has a negative coefficient of resistance; as the temperature rises, the carrier concentration increases and so the resistance falls.

(9) C

ⓔ $hf_0 = \phi = 1.9$ eV $= 1.9 \times 1.6 \times 10^{-19}$J, so $f_0 = \dfrac{1.9 \times 1.6 \times 10^{-19}\text{J}}{6.63 \times 10^{-34}\text{Js}} = 4.6 \times 10^{14}$ Hz

(10) A

ⓔ Using $E = hf = h\dfrac{c}{\lambda}$

energy of the incident photon $= \dfrac{(6.63 \times 10^{-34}\text{Js}) \times (3.00 \times 10^8\text{ms}^{-1})}{599 \times 10^{-9}\text{m}}$

$= 3.32 \times 10^{-19}$J $= \dfrac{3.32 \times 10^{-19}\text{J}}{1.6 \times 10^{-19}\text{JeV}^{-1}} = 2.08$ eV

Maximum KE = 2.08 eV − 1.90 eV = 0.18 eV

Question 11

The table below shows some features of the electromagnetic spectrum. Complete the table by filling in the missing entries labelled (a), (b), (c) and (d).

Radiation	Typical wavelength	Source
Visible light	(a)	Very hot objects
X-rays	(b)	(c)
(d)	3.0 cm	Very high frequency oscillator (klystron tube)

Total: 4 marks

ⓔ This question tests your knowledge of the properties of electromagnetic radiation. Full marks can be earned if the details in Table 1 on page 9 are known.

Student answer

(a) 400–700 nm ✓

(b) 10^{-14} m to 10^{-10} m ✓

(c) (electrons striking) tungsten anode ✓

(d) microwaves ✓

Question 12

Waves on the sea enter a harbour and are diffracted as they pass through the harbour gate.

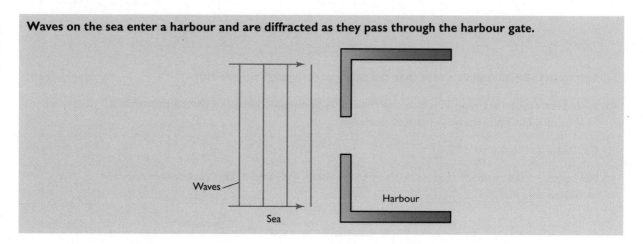

ⓔ This question tests your knowledge of diffraction and how wavefronts and wavelength are affected by obstructions and wave speed.

(a) Explain the meaning of diffraction. (1 mark)

(b) Copy the diagram, and add three more wavefronts of the wave entering the harbour. (2 marks)

ⓔ In part (b) a typical grade-C response may show circular wavefronts spreading out but have them closer together, farther apart or with variable wavelength after entering the harbour. Take care with these diagrams — sloppy efforts lose marks!

(c) Further up the coast, as the waves approach the shore, the wavelengths get shorter. What does this tell you about the speed of the waves in shallower water? (1 mark)

Total: 4 marks

ⓔ Part (c) requires just a 1-mark answer. Use the equation $v = f\lambda$ to relate speed and wavelength.

Student answer

(a) Diffraction is the spreading out of waves when they pass through an aperture or are obstructed by an object. ✓

(b) The drawing should show three wavefronts with a (semi-)circular shape having the same wavelength (i.e. spacing between consecutive wavefronts) as in the open sea ✓, and spreading out by at least 45° ✓.

(c) The waves slow down in the shallower water. ✓

Question 13

A 1.0 kΩ carbon resistor has a power rating of 100 mW.

ⓔ This question requires the use of the basic equations for electrical power and parallel circuits.

(a) Calculate the maximum current that can safely pass through the resistor. (1 mark)

(b) A student designs a circuit in which the resistor is connected between the terminals of a 12 V supply, but finds that the resistor burns out.

Explain why the resistor failed. (1 mark)

(c) His teacher recommends that he replace the resistor with four 1.0 kΩ resistors connected as shown below:

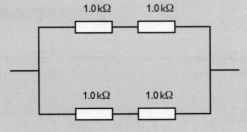

Show that the total resistance of the network is 1.0 kΩ. (1 mark)

Edexcel AS Physics

(d) Explain why the resistors in this network will not burn out when connected to the 12 V supply. (2 marks)

Total: 5 marks

Student answer

(a) Using $P = I^2R$, $I^2_{max} = \dfrac{P_{max}}{R} = \dfrac{100 \times 10^{-3}\,\text{W}}{1.0 \times 10^3\,\Omega} = 1 \times 10^{-4}\,\text{A}^2$, so $I_{max} = 1 \times 10^{-2}\,\text{A} = 10\,\text{mA}$ ✓

(b) When a voltage of 12 V is applied across a 1.0 kΩ resistor, a current of 12 mA flows. This is greater than the maximum current that the resistor can take, so it burns out. ✓

(c) The network is the same as two 2.0 kΩ resistors in parallel:
$\dfrac{1}{R} = \dfrac{1}{2.0 \times 10^3\,\Omega} + \dfrac{1}{2.0 \times 10^3\,\Omega} = 1 \times 10^{-3}\,\Omega \Rightarrow R = 1.0\,\text{kΩ}$ ✓

(d) The current in each arm of the network is $\dfrac{12\,\text{V}}{2\,\text{k}\Omega} = 6\,\text{mA}$ ✓

Each of the resistors therefore carries less than 10 mA, so they will not overheat. ✓

ⓔ A C-grade candidate may gain the marks for calculating the values of the current in parts (b) and (d) but lose the final mark if the values are not compared with the maximum value in part (a).

Question 14

A standing wave is set up in a string as shown in the diagram. The frequency of the supply is 60 Hz.

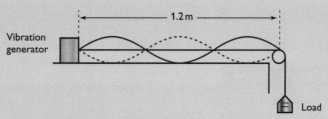

(a) Sketch the vibrating string, and label the position of one node and one antinode. (1 mark)

(b) What is the wavelength of the standing wave? (1 mark)

(c) The frequency of the vibration generator is altered until a standing wave with one more node is set up in the string. Calculate the frequency of the vibration generator when this occurs. (2 marks)

Total: 4 marks

ⓔ You can gain full marks for this question if the standard diagrams for standing waves (Figure 14) are learned.

Student answer

(a) A node should be labelled at one of the positions of minimum disturbance, and an antinode labelled at the midpoint of one of the loops. ✓

(b) There are one and a half wavelengths on the string: $1.5\lambda = 1.2\,\text{m}$, so $\lambda = 0.80\,\text{m}$ ✓

e A common mistake that candidates make with standing waves is to take the wavelength as the distance between two adjacent nodes or antinodes, instead of twice this length (i.e. the separation between every other node or every other antinode).

(c) With the addition of one node, there will now be two complete wavelengths along the string, so the wavelength will be $0.60\,\text{m}$ ✓
Using $v = f\lambda$, as the speed of the wave on the string is unchanged, we have
$$f_1\lambda_1 = f_2\lambda_2$$
and hence $f_2 = \dfrac{0.80\,\text{m} \times 60\,\text{Hz}}{0.60\,\text{m}} = 80\,\text{Hz}$ ✓

Question 15

A 60 W lamp gives out only 5% of its power as visible light.

(a) Show that the intensity of light at a distance of 1.5 m from the lamp is about 0.1 W m⁻². (2 marks)

(b) Calculate the distance from a 100 W lamp at which the intensity would be the same as that found above. (1 mark)

Total: 3 marks

e A simple application of the inverse square law is needed here.

Student answer

(a) $I = \dfrac{P}{4\pi d^2} = \dfrac{3.0\,\text{W}}{4\pi(1.5\,\text{m})^2}$ ✓
$= 0.11\,\text{W m}^{-2}$ ✓

e This is a 'show that' question, so the final result must be given to at least one more significant figure than what is stated in the question. Writing down the correct equation and then simply stating that the value is approximately $0.1\,\text{W m}^{-2}$ will cause you to lose the second mark.

(b) $0.11\,\text{W m}^{-2} = \dfrac{5.0\,\text{W}}{4\pi d^2} \Rightarrow d = \sqrt{\dfrac{5.0\,\text{W}}{4\pi \times 0.11\,\text{W m}^{-2}}} = 1.9\,\text{m}$ ✓

e The answer would be 2.0 m if you used the value $0.1\,\text{W m}^{-2}$ in the calculation; this would gain the second mark.

Question 16

The diagram shows some of the energy levels of a hydrogen atom.

Ionisation level ————————————— 0 eV

————————————— −1.51 eV

————————————— −3.41 eV

Ground state ————————————— −13.6 eV

ⓔ This question is about the absorption and emission of energy when electrons are raised and lowered between permitted 'energy levels'.

(a) Calculate the ionisation energy in joules for an electron in the −13.6 eV level.   (2 marks)

(b) A neutron of kinetic energy 12.9 eV collides with the atom. As a result, an electron in the ground state is raised to the −1.51 eV level. What is the kinetic energy of the neutron after the collision?   (1 mark)

ⓔ This mark will be lost if only the energy needed to raise the electron to the higher orbital is calculated. The question requires the determination of the residual energy of the neutron.

(c) A transition between which two energy levels will give rise to an emission of wavelength 654 nm?   (3 marks)

                                                                        Total: 6 marks

Student answer

(a) Ionisation energy $= 13.6\,\text{eV} \times 1.6 \times 10^{-19}\,\text{J eV}^{-1}$ ✓

$= 2.2 \times 10^{-18}\,\text{J}$ ✓

(b) Energy gained by electron $= -1.51\,\text{eV} - (-13.6\,\text{eV})$

$= 12.1\,\text{eV} =$ energy lost by neutron

Kinetic energy of neutron after collision $= 12.9\,\text{eV} - 12.1\,\text{eV} = 0.8\,\text{eV}$ ✓

(c) $E = hf = \dfrac{hc}{\lambda} = \dfrac{(6.63 \times 10^{-34}\,\text{J s}) \times (3.00 \times 10^{8}\,\text{m s}^{-1})}{654 \times 10^{-9}\,\text{m}}$ ✓

$= 3.04 \times 10^{-19}\,\text{J} = \dfrac{3.04 \times 10^{-19}\,\text{J}}{1.6 \times 10^{-19}\,\text{J eV}^{-1}} = 1.9\,\text{eV}$ ✓

This is the transition from the −1.51 eV level to the −3.41 eV level. ✓

ⓔ The final mark is lost if the transition is the wrong way round, or if the minus signs are omitted.

Question 17

In the early days of commercial radio, signals were transmitted on the continent to be received in Britain. The strength of the signals received often varied due to the interference of the signal transmitted directly to the aerial with that reflected off layers of the atmosphere.

Use the terms constructive interference, destructive interference and path difference to explain this phenomenon.

Total: 3 marks

(e) Although this question is asking for three definitions, it is essential to refer these to the radio waves. It is possible to correctly define the three terms and get no marks if no connection is made to the phenomenon described in the question. An A-grade candidate will link the path difference to the phase relationship as shown above.

Student answer

The part of the signal that has been reflected off the ionosphere has travelled further from transmitter to receiver than the direct signal. The difference between the lengths of the paths taken by the two signals is called the *path difference.* ✓

When the path difference is a whole number of wavelengths, the signals are in phase at the receiver, *constructive interference* takes place, and a stronger signal is received. ✓

When the path difference is an odd number of half-wavelengths, the waves arrive out of phase and *destructive interference* takes place, resulting in a weaker signal. ✓

Question 18

A light-dependent resistor (LDR) is used in a potential divider circuit that operates a light switch when the background illumination falls below a certain level.

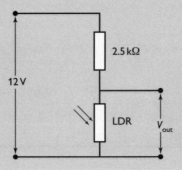

12 V

2.5 kΩ

LDR

V_{out}

(e) This question requires the application of the potential divider equation.

(a) In bright conditions the resistance of the LDR is about 500Ω. Calculate the output voltage in these conditions.   (2 marks)

The following graph shows that the resistance of the LDR increases when the illumination falls, so the output voltage of the potential divider will rise and eventually reach a level where it will activate a switch to turn on the lamp.

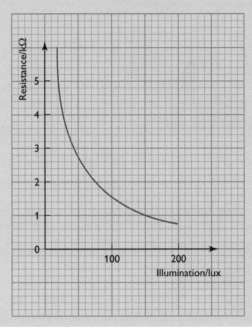

ⓔ You are expected to read points on a graph to ± one half of a division.

(b) The light switch is triggered when the output voltage reaches 4.5 V. Calculate the resistance of the LDR at the switching threshold, and use the graph to determine the background illumination when the lamp is turned on.   (3 marks)

Total: 5 marks

Student answer

(a) $V_{out} = \dfrac{0.50\,k\Omega}{2.5\,k\Omega + 0.50\,k\Omega} \times 12\,V$ ✓ $= 2.0\,V$ ✓

(b) $4.5\,V = \dfrac{R}{2.5\,k\Omega + R} \times 12\,V$ ✓

giving $R = 1.5\,k\Omega$ ✓

From the graph, the illumination corresponding to a resistance of 1.5 kΩ is 100 lux. ✓

ⓔ Answers between 95 and 105 lux are acceptable.

Question 19

A schematic diagram showing the use of an ultrasound A-scan to measure the diameter of a baby's head is given below. The graph represents the reflections from the front and rear of the skull as shown on a cathode ray oscilloscope monitor.

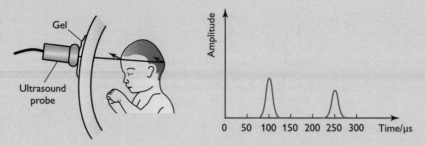

(a) What is ultrasound? (1 mark)

(b) An AS physics examination paper contained the question: 'Describe how an A-scan is used to measure the diameter of a fetal skull, and explain why a coupling gel is used between the probe and the skin.'

One student's answer was as follows:

Ultrasound is directed at the skull and reflected back to the probe. The reflections are shown as two peaks on the screen. The distance between the peaks represents the diameter of the skull.

The coupling gel is needed to replace the air between the probe and the skin because the ultrasound will reflect back off the air.

Discuss the student's answer, highlighting any incorrect or missing physics. (4 marks)

ⓔ The 'student's answer' given in the question illustrates many of the frequently encountered errors and omissions made by AS candidates. Such an answer would gain 0 marks from a possible 4.

(c) If the average speed of sound in soft tissue is $1560\,\text{ms}^{-1}$, use the information on the graph to determine the diameter of the baby's head. (3 marks)

ⓔ Remember to include the return distance in the pulse–echo calculation.

(d) The heartbeat of a fetus can be monitored using the Doppler effect. Describe how reflections of ultrasound from the surface of the beating heart are used to measure the heart rate. (4 marks)

Total: 12 marks

Edexcel AS Physics

ⓔ To gain full marks you will need to explain how the frequency of the reflected pulses changes when the surface of the heart moves towards and away from the detector.

Student answer

(a) Ultrasound is sound that has a frequency greater than that which can be recognised by the human ear (usually greater than 20 kHz). ✓

(b) *Pulses* of ultrasound are needed. ✓
The cathode ray oscilloscope shows the *time delay* between the reflection from the front of the skull and the reflection from the rear. ✓
The distance is then calculated as speed × time. ✓
The time measured is the time needed for the pulse to travel to the rear of the skull *and back again*. So the depth of the skull is $\frac{vt}{2}$ ✓

ⓔ Any three of the statements given in part (b) would earn 3 marks, plus one of the following for the fourth mark:
- The ultrasound does not reflect off the air; it is almost totally reflected at the air–skin boundary. ✓
- There is a large difference in acoustic impedance between air and soft tissue. The acoustic impedance of the gel is similar to that of skin, thus allowing less of the ultrasound to be reflected and more to be transmitted. ✓

    The most common omission is the mention of 'pulses' — a continuous stream of ultrasound will mix up all the reflections and show nothing.

(c) Using speed = $\frac{\text{distance}}{\text{time}}$ or distance = speed × time ✓

we get $d = \frac{vt}{2} = \frac{(1560\,\mathrm{m\,s^{-1}}) \times (150 \times 10^{-6}\,\mathrm{s})}{2}$ ✓ = 0.12 m ✓

(d) The Doppler effect refers to the change in the frequency of a signal that is observed when the source is moving relative to the observer. ✓

Pulses of ultrasound are directed at the heart and reflected back to the probe. ✓

The frequency of the reflected waves is compared with the frequency of the transmitted waves, and the change in frequency is displayed. ✓

When the heart wall is moving towards the probe, a frequency increase will be observed; and as the heart wall moves away, the frequency will be observed to decrease. ✓

The number of frequency variations per minute gives the heart rate. ✓

ⓔ Any four of the above will earn the 4 marks.

Question 20

On the diagram below, graph X shows how the potential difference across the terminals of a cell depends on the current in the cell. Graph Y is the voltage–current characteristic for a filament lamp.

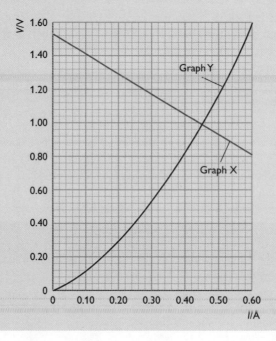

ⓔ This question examines the relationship between the emf, internal resistance and terminal potential difference for a cell (see the equation on p. 42) and the variable resistance of a filament lamp at different temperatures. Note that the axes of the graph are the opposite of those normally used for the characteristics of a lamp (p. 37).

(a) As the current increases, what can be deduced from the graphs about:
   (i) the internal resistance of the cell?
   (ii) the resistance of the filament lamp? (2 marks)

(b) Use graph X to determine:
   (i) the emf of the cell
   (ii) the internal resistance of the cell (2 marks)

ⓔ The question requires that information is to be determined from the graph. Candidates who try to calculate the values from other data will gain no marks.

(c) When the lamp is connected to the cell, what is:

   (i) the current in the lamp?

   (ii) the resistance of the lamp?

   (iii) the power developed in the lamp?   (3 marks)

(d) Draw a circuit diagram of the circuit you would set up to obtain the data for graph **Y** using two cells connected in series, which are identical to the one in graph **X**, for the power supply.   (3 marks)

Ⓔ The circuit must include the appropriate meters, correctly connected, and some means of varying the current or potential difference.

(e) Explain:

   (i) why you would need two cells in series, rather than a single cell, to achieve the results shown by graph **Y**

   (ii) how you would obtain the data.   (3 marks)

                                                                            Total: 13 marks

Ⓔ You will need to consider the effect of the internal resistance of the cell on the maximum terminal potential difference.

Student answer

(a) (i) The internal resistance is constant. ✓

   (ii) The resistance of the lamp rises as the voltage or current increases. ✓

(b) Graph X represents the relationship $V = \mathcal{E} - Ir$.

   (i) The emf $\mathcal{E}$ is the intercept on the V-axis, i.e. 1.52 V ✓

   (ii) The internal resistance r is the negative of the gradient, i.e. $= -\dfrac{(0.80 - 1.52)\,\text{V}}{0.60\,\text{A}}$

   $1.2\,\Omega$ ✓

(c) When the lamp is connected to the cell, the values of V and I will correspond to the coordinates of the point of intersection between graph X and graph Y.

   (i) $I = 0.45\,\text{A}$ ✓

   (ii) $V = 0.98\,\text{V}$, so $R = \dfrac{V}{I} = \dfrac{0.98\,\text{V}}{0.45\,\text{A}} = 2.2\,\Omega$ ✓

   (iii) $P = VI = 0.44\,\text{W}$ ✓

(d) The diagram should show:

   a battery of two cells ✓

   a means of varying the voltage (e.g. a potential divider) ✓

   an ammeter in series and a voltmeter in parallel with the lamp ✓

(e) (i) With a single cell connected to the lamp, the maximum terminal potential difference would be less than 1.6 V. So, to cover the full range (up to about 1.6 V), a second cell is needed. ✓

   (ii) Adjust the potentiometer or variable resistor to give a range of voltages (e.g. 0 to 1.60 V at 0.20 V intervals) ✓, and read off the corresponding currents from the ammeter ✓.

Question 21

A refractometer is used in the food industry to determine the sugar concentration in fruit juices by measuring the refractive index of the juices and comparing these with the values for standard sugar solutions.

A simple form of refractometer is shown in the diagram below.

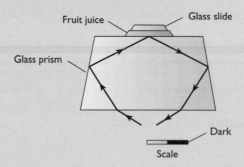

A ray of light is passed through the prism as shown, and the critical angle is measured using the position of the light–dark boundary on the scale.

(e) You will need to use the expressions linking the critical angle to the refractive index of the media on each side of the boundary (p. 23)

(a) Explain the meaning of critical angle, and mark its position on a sketch of the refractometer.   (3 marks)

(b) Before the juice is added, the critical angle for the glass–air boundary is measured. If this value is 41.0°, calculate the refractive index of the glass.   (1 mark)

(c) When fruit juice is used, the critical angle at the glass–juice interface is 64.0°. Determine the refractive index of the juice.   (2 marks)

(d) The instrument has been calibrated using sugar solutions of different concentrations. The results are given in the table.

Concentration of sugar solution (%)	Refractive index of sugar solution
0	1.33
20	1.36
40	1.40
60	1.45

Plot a graph of these results.   (4 marks)

(e) Marks are awarded for correctly labelled axes, appropriate choice of scale, accurate plotting (± half a division) and drawing the line of best fit. In this case a candidate would lose 2 marks if both scales started at the origin (so that the plots only covered a small region of the graph) and no % was included in the concentration region.

(e) Use the graph to determine the sugar concentration of the fruit juice.   (1 mark)

Total: 11 marks

Student answer

(a) The critical angle is the angle of incidence in the denser medium ✓ for which the angle of refraction in the less dense medium is 90°.✓

The critical angle should be drawn between the incident ray and the normal at the glass–juice interface. ✓

(b) Refractive index of glass $= \dfrac{1}{\sin 41°} = 1.52$ ✓

(c) $\sin 64° = \dfrac{\text{refractive index of juice}}{\text{refractive index of glass}}$ ✓

hence $\mu_{\text{juice}} = 1.52 \sin 64° = 1.37$ ✓

(d) The graph should have the following attributes:
- y-axis labelled 'refractive index' and restricted to a range from 1.30 to 1.50 ✓
- x-axis labelled 'concentration' with a range between 0 and 60% (100% OK) ✓
- four points plotted according to the data in the table ✓
- curve drawn through the points – must be a curve, not a straight line ✓

(e) Sugar concentration of juice $= 23 \pm 1\%$

ⓔ A grade-A candidate should realise that the graph is a gentle curve; a grade-C candidate may try to force a straight line through the points, and hence lose a mark in part (d), but could still gain the mark in part (e) if the estimated concentration is within those limits.

Test Paper 2

Questions 1–10

For questions 1–10 select one answer from A to D.

(1) **Blue light is deviated more than red light when it enters a glass block, from air, at an angle of incidence less than 90°. This is because:**

   A   it has a longer wavelength

   B   it has a lower frequency

   C   it travels at a greater speed in glass than does red light

   D   it travels at a lower speed in glass than does red light   (1 mark)

(2) **When a guitar string is plucked, a wave on the string and a sound wave in air are produced. Which of the following statements is true?**

   A   The sound wave in air is longitudinal and stationary.

   B   The sound wave in air is transverse and progressive.

   C   The wave on the string is longitudinal and stationary.

   D   The wave on the string is transverse and stationary.   (1 mark)

Questions 3 and 4 relate to the resistor network shown below:

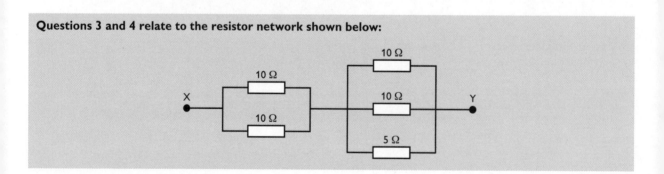

(3) The total resistance of the network is

A 7.5 Ω

B 12.5 Ω

C 22.5 Ω

D 45 Ω   (1 mark)

(4) When a potential difference of 12 V is applied across X and Y, the current in the 5.0 Ω resistor is

A 0.16 A

B 0.80 A

C 1.6 A

D 4.8 A   (1 mark)

ⓔ Use the potential divider equation to determine the potential difference across the second set of resistors. This will be the p.d. across the 5 Ω resistor; hence the current can be calculated.

(5) A student connects two loudspeakers to the same terminals of a signal generator. The speakers are placed about 1 metre apart in the school hall. A fellow student walks along a line parallel to the speakers, and hears a series of loud and quiet sounds. If there is a quiet sound at a point equidistant from both speakers, this is because:

A   one speaker is much louder than the other

B   the speakers are in antiphase

C   there is a path difference of half a wavelength between the student and the speakers

D   there is no path difference between the student and the speakers   (1 mark)

(6) Ultrasound is preferred to X-rays for some diagnostic images because:

A   it gives a more detailed image

B   it is a longitudinal wave

C   it is less harmful to the patient

D   it penetrates the body more easily   (1 mark)

In questions 7 and 8, which of the following graphs best represents the quantities described when they are plotted on the y- and x-axes? Each graph may be used once, more than once or not at all.

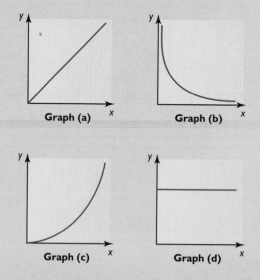

(7) y-axis: the intensity of light from a filament lamp; x-axis: the distance from the lamp

   A   graph (a)

   B   graph (b)

   C   graph (c)

   D   graph (d)

(1 mark)

ⓔ You should be aware that the intensity (radiant flux) obeys an inverse square law.

(8) y-axis: the resistivity of a fixed length of iron wire at constant temperature
x-axis: the diameter of the wire

   A   graph (a)

   B   graph (b)

   C   graph (c)

   D   graph (d)

(1 mark)

ⓔ Beware! The question relates to *resistivity* NOT resistance.

Questions 9 and 10 relate to the following diagram, which shows some of the energy levels for a mercury atom.

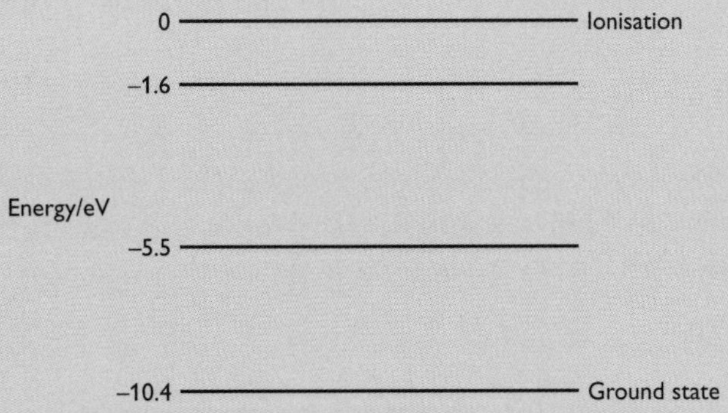

```
         0  ——————————————————————————  Ionisation
      −1.6  ————————————————————————

 Energy/eV
      −5.5  ————————————————————————


     −10.4  ————————————————————————  Ground state
```

(9) The ionisation energy for an electron in the ground state is

A 2.6×10^{-19} J

B 7.8×10^{-19} J

C 8.8×10^{-19} J

D 1.7×10^{-18} J   (1 mark)

ⓔ To ionise the atom the electron must be displaced from the ground state to the ionisation level.

(10) The wavelength of the photon emitted when an electron falls from the −1.6 eV level to the −5.5 eV level is:

A   120 nm

B   230 nm

C   250 nm

D   320 nm   (1 mark)

Total: 10 marks

ⓔ Convert the change in energy level to joules; use $\Delta E = hf$ and $c = f\lambda$.

Answers

(1) D

ⓔ Refractive index is the ratio of the speed in air to the speed in glass; blue light travels more slowly than red in glass, so its refractive index is larger, leading to greater deviation.

(2) D

ⓔ A standing (i.e. stationary) wave is set up on the string; waves on strings are transverse, whereas sound waves are longitudinal and progressive.

(3) A

ⓔ This network consists of two parallel circuits connected in series. The resistance of the parallel pair on the left-hand side is $5.0\,\Omega$, and the resistance of the three-resistor circuit on the right-hand side is $2.5\,\Omega$. The total resistance is therefore $7.5\,\Omega$.

(4) B

ⓔ There will be 8.0 V across the first pair of resistors and 4.0 V across the other three, (including the $5\,\Omega$ resistor), so $I = \dfrac{4.0\,\text{V}}{5.0\,\Omega} = 0.80\,\text{A}$.

(5) B

ⓔ As the point is equidistant from the two speakers, the path difference is zero; as the sound is quiet, there must be destructive interference at that point, which means that the waves are in antiphase. Therefore the speakers must be in antiphase.

(6) C

ⓔ Ultrasound is non-ionising radiation, and so does not cause tissue damage in the patient.

(7) B

ⓔ The intensity falls off according to an inverse square law, and so the graph is of the form $y = k/x^2$.

(8) D

ⓔ Provided the temperature is not changed, the resistivity of a metal is constant. While the resistance of a wire will be affected by the diameter, the resistivity is a property of the material, and is independent of the dimensions.

(9) D

ⓔ Energy needed for ionisation is 10.4 eV.

$10.4\,\text{eV} = 10.4\,\text{eV} \times (1.6 \times 10^{-19}\,\text{JeV}^{-1}) = 1.7 \times 10^{-18}\,\text{J}$

(10) D

ⓔ $\lambda = \dfrac{hc}{E} = \dfrac{6.63 \times 10^{-34}\,\text{Js} \times 3.00 \times 10^{8}\,\text{ms}^{-1}}{[-1.6 - (-5.5)]\text{eV} \times (1.6 \times 10^{-19}\,\text{JeV}^{-1})} = 3.2 \times 10^{-7}\,\text{m} = 320\,\text{nm}$

Question 11

An electric motor drives a pulley wheel with a length of string attached so that it can raise a load of mass 200 g. When the motor is operating, the voltage across its terminals is 5.0 V and it carries a current of 500 mA.

(a) Calculate the input power supplied to the motor.   (1 mark)

(b) The 200 g load is raised by 0.50 m in 0.80 s. Calculate the useful power output of the motor, and hence determine its efficiency.   (3 marks)

Total: 4 marks

ⓔ Remember to include g in the calculation of the power output.

Student answer

(a) $P = IV = 0.500\,\text{A} \times 5.0\,\text{V} = 2.5\,\text{W}$

(b) Useful power output $= \dfrac{\text{work done}}{\text{time}} = \dfrac{0.200\,\text{kg} \times 9.8\,\text{m s}^{-2} \times 0.50\,\text{m}}{0.80\,\text{s}} = 1.2\,\text{W}$ ✓✓

Efficiency $= \dfrac{\text{useful power output}}{\text{total power input}} \times 100\% = \dfrac{1.2\,\text{W}}{2.5\,\text{W}} \times 100\% = 48\%$ ✓

Question 12

The diagram shows a series of wavefronts moving on the surface of water in a ripple tank.

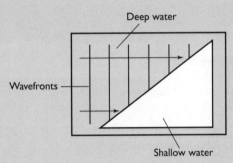

(a) Explain the meaning of wavefront.   (1 mark)

(b) The wavefronts move into a region of shallow water, where the wave speed is reduced.

Copy the diagram and add the positions of the wavefronts as they travel through the shallow water.   (3 marks)

Total: 4 marks

ⓔ This question examines the the principle of refraction and variation in wavelength when the wave speed changes. It is important to take care when drawing wavefront diagrams. Marks can be lost for sloppy work such as non-parallel lines or variable spacing between successive lines.

Student answer

(a) A wavefront is a line joining points in a wave that are in phase. ✓

(b) The diagram should show:
- at least three parallel wavefronts ✓
- the wavefronts changing direction, slanting to the left so that the direction of propagation (which is perpendicular to the wavefronts) bends downward towards the normal to the deep–shallow water boundary ✓
- the wavelength (i.e. spacing between consecutive wavefronts) becoming shorter ✓

Question 13

Electrons fired from an electron gun in an evacuated tube strike a thin graphite crystal. On a fluorescent screen beyond the crystal, a pattern of concentric rings is observed.

(a) If the accelerating voltage across the electron gun is 4.50 kV, calculate the maximum kinetic energy of the electrons leaving the gun.

(1 mark)

(b) How does this apparatus demonstrate the wave nature of electrons, and what can be deduced about their wavelength?

(2 marks)

Total: 3 marks

ⓔ This question is about how the degree of diffraction is dependent on the size of the aperture.

Student answer

(a) Kinetic energy gained by electron = work done on electron = $V \times e$
$$= (4.50 \times 10^3\,\text{V}) \times (1.6 \times 10^{-19}\,\text{C}) = 7.2 \times 10^{-16}\,\text{J} \checkmark$$

(b) The electrons are diffracted (the concentric rings are a diffraction pattern), so they must have wave properties. ✓

The wavelength of the electrons must be similar to the atomic separation in the graphite crystal. ✓

Question 14

A 10 kΩ potentiometer is used to provide a continuously variable power supply up to a maximum voltage of 5.0 V.

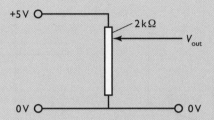

The potentiometer is adjusted so that the upper section has a resistance of 2.0 kΩ.

(a) Calculate the output voltage.   (1 mark)

(b) An 8.0 kΩ resistor is connected across the output terminals without altering the setting of the potentiometer.

What will be the value of the output voltage with the resistor connected?   (2 marks)

Total: 3 marks

ⓔ It is quite common for potential divider questions to ask about the effect of putting a load resistor across the output, sometimes in the form of a voltmeter. It should be realised that, unless the added resistor has very large resistance in comparison with the output resistance of the potential divider, the voltage output will get smaller.

Student answer

(a) The resistance in the lower section is 8.0 kΩ, so $V_{out} = \dfrac{8.0\,k\Omega}{10\,k\Omega} \times 5.0\,V = 4.0\,V$ ✓

(b) Total resistance between the output terminals is that of two 8.0 kΩ resistors in parallel, i.e. 4.0 kΩ ✓

$V_{out} = \dfrac{4.0\,k\Omega}{2.0\,k\Omega + 4.0\,k\Omega} \times 5.0\,V = 3.3\,V$ ✓

Question 15

A child dips a wire frame into soap solution to blow some bubbles. She notices that there are multicoloured patterns on the film when it is held up to the light.

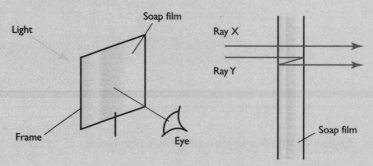

The second diagram shows a cross-section of the soap film with two adjacent rays passing through a certain region. Ray X passes straight through the film, but ray Y undergoes two reflections on the inner surfaces before emerging. In this region the thickness of the film is about 450 nm. The wavelength of blue light in the soap solution is about 300 nm.

ⓔ You need to be aware that interference will take place between the two rays. An A-grade student will be able to relate the path difference and the wavelength when constructive interference occurs (see p. 15).

(a) Explain why this region appears blue to the child. (3 marks)

(b) The soap film increases in thickness from top to bottom. Suggest a reason for this. (1 mark)

(c) When a red lamp is viewed through the film, a series of bright and dark horizontal stripes is seen. Explain this effect, and determine the minimum thickness of the soap film for a dark line to appear if the wavelength of red light in the soap solution is 500 nm. (3 marks)

Total: 7 marks

Student answer

(a) The reflected ray travels back and forth between the surfaces. The path difference between this ray and the one passing straight through is therefore 2 × 450 nm = 900 nm. ✓

The path difference is equal to three complete wavelengths of blue light in the solution. ✓

So the two waves emerge in phase and constructive superposition occurs for blue light. ✓

(b) Gravitational force pulls the liquid down; or: the solution flows down, making the film thicker at the bottom. ✓

Edexcel AS Physics

(c) When the path difference is a whole number of wavelengths, constructive interference occurs; when the path length differs by an odd number of half-wavelengths, destructive interference takes place. ✓

The minimum thickness t for destructive interference is the thickness for which $2t = \lambda/2$ ✓, that is, $t = \lambda/4 = 125\,nm$ ✓.

ⓔ Producing destructive interference by means of two reflections that give rise to a path difference of half a wavelength is commonly seen in examination questions. Examples include reflections from the edges of the 'bumps' on CDs and non-reflective coatings on lenses.

Question 16

Under steady light conditions, the emf of a photovoltaic cell is measured by connecting a voltmeter of very high resistance directly across the cell's terminals. The emf of a particular cell is found to be 1.60 V.

When a 100 Ω resistor is connected between the terminals of this cell, the potential difference across it is 1.54 V.

(a) Show that the internal resistance of the cell is about 4 Ω.   (2 marks)

ⓔ Remember that because this is a 'show that' question, the result must be given to one significant figure more than the value stated in the question.

A student wishes to investigate how the maximum power output of the cell depends on the intensity of the light falling on it. He has read that the maximum output power will be achieved when the external resistance is equal to the internal resistance, so he replaces the 100 Ω resistor with one having the same value as the internal resistance of the cell.

(b) What will be the value of the voltage across the resistor if the light conditions are unchanged? (2 marks)

(c) Describe how the student would perform the investigation. You will be expected to state what additional equipment is required, draw a circuit diagram and describe the measurements that need to be taken.   (5 marks)

ⓔ Descriptions of standard experiments are often given a 'quality of written communication' (QWC) mark. Answers need to be grammatically correct, with few spelling mistakes, and must contain the appropriate technical and scientific terminology. A series of phrases or bullet points is acceptable.

(d) What assumption must the student make regarding the internal resistance of the cell?   (1 mark)

Total: 10 marks

Student answer

(a) $I = \dfrac{V}{R} = \dfrac{1.54\,V}{100\,\Omega} = 0.0154\,A$ ✓

From $\mathcal{E} = V + Ir$ we have $r = \dfrac{\mathcal{E}-V}{I} = \dfrac{1.60\,V - 1.54\,V}{0.0154\,A} = 3.9\,\Omega$ ✓

(b) When $R = r = 4\,\Omega$, $I = \dfrac{1.60\,V}{4.0\,\Omega + 4.0\,\Omega} = 0.20\,A$ ✓ (or 0.21 A if $R = 3.9\,\Omega$ used)

so voltage across the resistor is $0.20\,A \times 4.0\,\Omega = 0.80\,V$ ✓

ⓔ When the external resistance equals the internal resistance of the cell, the terminal voltage will equal half the value of the emf, as the same work has to be done in passing the current through the cell as in the external circuit.

(c) Additional equipment: ammeter and light-meter. ✓

Circuit diagram should show ammeter in series and voltmeter in parallel with the resistor. ✓

For a range of light intensities ✓ measure the corresponding values of I and V ✓ so that the power can be calculated using $P = IV$ ✓.

(d) It must be assumed that the internal resistance of the cell remains the same for all light conditions. ✓

Question 17

A recorder can be modelled as a tube, open at each end, in which standing waves may be set up.

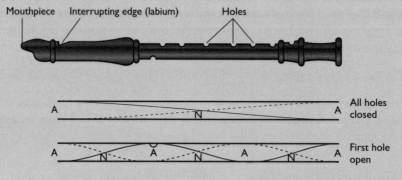

Mouthpiece   Interrupting edge (labium)   Holes

All holes closed

First hole open

The effective length of the recorder is 48.5 cm, and the fundamental frequency when all holes are closed is 349 Hz.

ⓔ This question requires you to deduce the wavelengths of standing waves in a pipe and use the wave equation to determine wave speed or frequency.

(a) **Use the information given above to determine a value for the speed of sound in air.** (2 marks)

(b) **Calculate the frequency of the note produced when the first hole is open.** (2 marks)

(c) **At higher temperatures, the speed of sound in air increases. What effect will an increase in the air temperature have on the pitch of a note played on a recorder?** (1 mark)

Total: 5 marks

Student answer

(a) The wavelength λ is $2 \times 0.485\,\text{m} = 0.970\,\text{m}$ ✓

$v = f\lambda = 349\,\text{Hz} \times 0.970\,\text{m} = 339\,\text{m s}^{-1}$ ✓

(b) Length of recorder = 1.5 wavelengths, so $v = \dfrac{0.485\,\text{m}}{1.5} = 0.323\,\text{m}$ ✓

$f = \dfrac{v}{\lambda} = \dfrac{339\,\text{m s}^{-1}}{0.323\,\text{m}} = 1050\,\text{Hz}$ (to 3 s.f.) ✓

ⓔ An A-grade candidate may use the fact that the wavelength is one-third that of the fundamental note to deduce that the frequency will be three times as large: $f = 3 \times 349\,\text{Hz} = 1047\,\text{Hz}$.

(c) When the temperature rises, the speed of sound increases. Because $v = f\lambda$ and the wavelength λ is unchanged, the frequency of the note must increase and so the pitch gets higher. ✓

Question 18

Some transparent materials are optically active and can rotate the plane of polarisation of light passing through them.

(a) **Explain the meaning of plane of polarisation.** (1 mark)

A polarimeter is a device for measuring the rotation of polarised light after it has passed through an optically active liquid. A simple version is shown below. It consists of a light source beneath a fixed polarising film (the polariser), together with a second, similar sheet of Polaroid that can be rotated around an angular scale (the analyser). A glass container holding a fixed volume of an optically active liquid is placed between the polariser and the analyser.

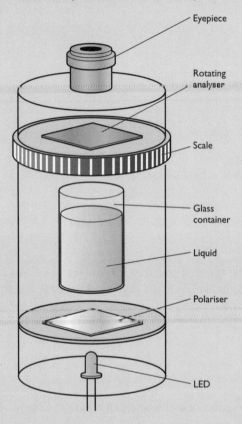

(b) Describe a method that you could use to measure the angle of rotation for a given sample of the liquid.

(3 marks)

A low-calorie drink contains artificial sweeteners that are optically active. When a sample of the drink is placed in the polarimeter, the angle of rotation is found to be 40°.

The polarimeter is calibrated by measuring the angle of rotation for a range of concentrations of the sweetener. The results are given in the following table.

Concentration/%	Angle of rotation/°
10	15
15	30
20	46
25	60

(c) Plot a graph of angle of rotation against concentration, and use this to determine the concentration of sweetener in the drink.

(4 marks)

🅔 Marks are awarded for correctly labelled axes, appropriate choice of scale, accurate plotting (± half a division) and drawing the line of best fit. Lines drawn on the graph (using a ruler) from the appropriate angle to the line and from there to the concentration axis will reduce the possibility of error in the final answer.

(d) Give one further application of optical activity.   (1 mark)

Total: 9 marks

Student answer

(a) The plane of polarisation is the plane in which the vibrations of a (transverse) wave take place. ✓

(b) Without any solution in the container, adjust the analyser until the light is blocked out. ✓

Take the scale reading (it should be zero). ✓

With a sample in place, rotate the analyser again until the light vanishes. Note the scale reading and so determine the angle through which the plane of polarisation has been rotated. ✓

(c) The graph should have:
- the axes labelled: concentration/% and rotation/° ✓
- suitable scales: e.g. 0–30% and 0–80° ✓
- four points plotted and a straight line of best fit drawn ✓

Concentration of sweetener in the sample is estimated to be 18–19% ✓

🅔 A similar marking scheme is followed for most graphs. It is best to copy the labels for the axes directly from the given table (e.g. 'concentration/%') so that the units are always included. The scale should be such that the given range of values fills most of the available space. Sometimes marks are awarded separately for the plotted points and the line or curve through them (especially when a curve is expected); but in this question, the points and the line both need to be drawn accurately to gain the 1 mark.

(d) Stress patterns in Perspex models, liquid crystal displays etc. ✓

🅔 A different example is needed. A candidate who gives another example of liquid concentration would not gain this mark.

Question 19

A photocell is a device that can be used to measure the intensity of light incident upon it.

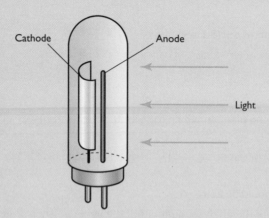

When photons of light hit the cathode, some electrons may be released. This emission depends on the energy of the photon and the work function of the cathode.

> ℯ This question requires the understanding and use of Einstein's photoelectric equation (see p. 47)

(a) Explain the meaning of work function, and state the conditions under which photoelectric emission will take place. **(2 marks)**

(b) Caesium has a work function of 1.9 eV. Calculate the lowest frequency of radiation that will produce photoelectric emission from a caesium-coated cathode. **(3 marks)**

(c) Visible light has a range of wavelengths from 400 nm to 700 nm. Explain, with appropriate calculations, whether a photocell with a caesium-coated cathode would be suitable for the whole of the visible range. **(3 marks)**

> ℯ 'Appropriate calculations' means that both extreme frequencies must be calculated and compared with the threshold frequency.

(d) A caesium cathode is exposed to the full range of wavelengths of the visible spectrum. Show that the maximum kinetic energy of the photoelectrons is about 1 eV. **(3 marks)**

> ℯ Because this is a 'show that' question, the result (and the calculation leading up to it) must be given to two significant figures since the value stated in the question only has one significant figure.

(e) Give one reason why some photoelectrons will be emitted with less than 1 eV of kinetic energy. **(1 mark)**

Total: 12 marks

Student answer

(a) The work function of a material is the minimum energy that is needed to remove an electron from atoms close to the surface. ✓

Photoelectric emission will take place if the energy of an incident photon is equal to or greater than the work function. ✓

(b) $1.9\,\text{eV} = 1.9 \times 1.6 \times 10^{-19}\,\text{J}$ ✓ $= 3.04 \times 10^{-19}\,\text{J}$

Using $\phi = hf_0$, we have $3.04 \times 10^{-19}\,\text{J} = 6.63 \times 10^{-34}\,\text{Js} \times f_0$ ✓

so $f_0 = 4.6 \times 10^{14}\,\text{Hz}$ ✓

(c) For red light: $f = \dfrac{c}{\lambda} = \dfrac{3.00 \times 10^8\,\text{m s}^{-1}}{700 \times 10^{-9}\,\text{m}} = 4.3 \times 10^{14}\,\text{Hz}$ ✓

For blue light: $f = \dfrac{3.00 \times 10^8\,\text{m s}^{-1}}{400 \times 10^{-9}\,\text{m}} = 7.5 \times 10^{14}\,\text{Hz}$ ✓

A photocell with a caesium-coated cathode would not cover the whole of the visible range, because visible light with frequencies from $4.3 \times 10^{14}\,\text{Hz}$ to $4.6 \times 10^{14}\,\text{Hz}$ will not be able to release a photoelectron. ✓

ⓔ The first mark will be given for using the $f = \frac{c}{\lambda}$ equation with values correctly substituted for either the longest or the shortest wavelength, but correct values for both of the frequencies are needed to gain the second mark.

(d) Maximum photon energy $= hf_{max} = (6.63 \times 10^{-34}\,\text{J s}) \times (7.5 \times 10^{14}\,\text{Hz})$ ✓

$$= 4.97 \times 10^{-19}\,\text{J} = \dfrac{4.97 \times 10^{-19}}{1.6 \times 10^{-19}}\,\text{eV} = 3.1\,\text{eV}$$ ✓

$hf_{max} = \phi + KE_{max} \Rightarrow KE_{max} = 3.1\,\text{eV} - 1.9\,\text{eV} = 1.2\,\text{eV}$ ✓

(e) Some of the electrons that were further from the surface may transfer some energy in the process of reaching the surface. The less energetic photons in the visible light range (those associated with the longer wavelengths towards the red end of the spectrum) will release electrons with lower kinetic energy. ✓

ⓔ Either of these reasons would gain the mark.

Question 20

In metallic conductors an electric current is a flow of free electrons.

(a) What are free electrons? (1 mark)

(b) At normal room temperatures, the free electrons in a length of copper wire that is not connected to a power supply have an average velocity of about 500 m s⁻¹, which is due to their thermal energy. Why does no current flow in the wire? (1 mark)

(c) When the copper wire is part of a circuit connected to a battery, the current flowing in the wire can be represented by the equation $I = nqvA$, where A is the area of cross-section of the wire and q is the charge carried by an electron ($1.6 \times 10^{-19}\,\text{C}$). Explain the meanings of n and v in the equation. (2 marks)

ⓔ The question requires an *explanation* of the terms; simply stating 'carrier concentration' and 'drift velocity' will gain no marks.

(d) Show that the value of *v* is about 0.5 mm s⁻¹ in a length of copper wire of cross-sectional area 0.085 mm² which is carrying a current of 0.50 A, given that *n* is 8.0 × 10²⁸ m⁻³. Comment on this value.

(3 marks)

(e)

Material	Resistivity/Ω m
Copper	1.7×10^{-8}
Constantan	4.9×10^{-7}
Silicon	2.4×10^{-3}

Use the equation *I* = *nqvA* to explain why:
(i) silicon has a much greater resistivity than copper
(ii) the resistivity of copper increases with increasing temperature while that of silicon decreases with temperature

(3 marks)

ⓔ The question asks you to 'use the equation', so reference must be made to drift velocity in the first case and to carrier concentration in the second. Many candidates give good explanations without reference to the equation and thus lose the marks.

(f) A technician wishes to make a 1.0 Ω resistor. She has a reel of copper wire of cross-sectional area 0.085 mm².
(i) What length of wire will she need to make the resistor?
(ii) Why would it be better to use a length of constantan wire of the same diameter to make the resistor?

(3 marks)

Total: 13 marks

Student answer

(a) Free electrons are electrons that do not occupy a fixed position in the atomic structure and are able to move throughout the lattice. ✓

(b) The motion of the electrons is random, so there is no net transfer of charge. ✓

(c) *n* is the carrier concentration — the number of charge carriers (electrons) per unit volume. ✓

v is the drift velocity of the carriers — the speed at which they move along the conductor when a current flows. ✓

(d) $v = \dfrac{I}{nqA} = \dfrac{0.50\,\text{A}}{(8.0 \times 10^{28}\,\text{m}^{-3}) \times (1.6 \times 10^{-19}\,\text{C}) \times (0.085 \times 10^{-6}\,\text{m}^2)} = 4.6 \times 10^{-4}\,\text{m s}^{-1}$

$= 0.46\,\text{mm s}^{-1}$ ✓✓

This velocity is very small. ✓

(e) (i) Silicon has a much lower carrier concentration than copper and so will have a greater resistivity. ✓

(ii) In copper, as the temperature rises the atomic vibrations increase, and so the free electrons undergo more collisions with the lattice; the drift velocity is therefore lowered and the resistivity will rise. ✓ In silicon, an increase in temperature releases many more charge carriers, so n becomes much larger, increasing the flow of charge and hence reducing the resistivity. ✓

(f) (i) $R = \dfrac{\rho l}{A} \Rightarrow l = \dfrac{RA}{\rho}$ ✓ $= \dfrac{1.00\,\Omega \times (0.085 \times 10^{-6}\,m^2)}{1.7 \times 10^{-8}\,m^{-3}} = 5.0\,m$ ✓

(ii) Because its resistivity is greater, a shorter length of constantan is needed to make the resistor. ✓

Knowledge check answers

1 (a) microwaves, (b) ultraviolet, (c) radio
2 (a) visible, (b) infrared, (c) X-rays or gamma radiation
3 Coherent sources of similar amplitude.
4 Boxes labelled microphone, amplifier, (electronic) inverter and headphones, connected in order with arrows.
5 $\lambda = v/f = 1.33\,\text{m}$
6 $\sin\theta = 1.48/1.56 = 0.949 \quad \theta = 72°$
7 Sound waves are longitudinal. Longitudinal waves cannot be plane polarised because the oscillations are always parallel to the direction of wave motion.
8 The wavelength of mobile phone signals is much shorter than radio waves, so there will be less diffraction (spreading of the waves around the obstruction).
9 If a continuous wave were used, all the reflections from all surfaces would be received at the same time, making it impossible to distinguish specific echoes.
10 As the car approaches, there is an increase in frequency of 20 Hz, so, when the car moves away from the observer at the same speed, there will be a decrease of 20 Hz. The observed frequency will be 80 Hz.
11 Three 15 Ω resistors in parallel are equivalent to a single 5 Ω resistor, and two 18 Ω resistors behave like a 9 Ω resistor. The total resistance is therefore 14 Ω.
12 $VIt = 230\,\text{V} \times \dfrac{230\,\text{V}}{20\,\Omega} \times 300\,\text{s} = 790\,\text{kJ}$

13 $\rho = 1.8\,\Omega \times \dfrac{\pi(0.25 \times 10^{-3}\,\text{m})^2}{2.00\,\text{m}} = 1.8 \times 10^{-7}\,\Omega\text{m}$
14 $V = \dfrac{0.40\,\text{m} \times 12\,\text{V}}{1.00\,\text{m}} = 4.8\,\text{V}$
15 The two cells in parallel have an emf of 2.0 V and an internal resistance of 0.25 Ω. When combined in series with the other cell the total emf will be 4.0 V and the total internal resistance 0.75 Ω.
16 Using the equation $I = nqvA$, it can be seen that when I, q and A are constant, the drift velocity is inversely proportional to the carrier concentration. The carrier concentration is much larger in conductors so the drift velocity will be much less than in semiconductors.
17 The resistivity of a material with a positive coefficient will increase when its temperature is increased, whereas that of a material having a negative coefficient of resistivity will decrease as it temperature rises.
18 Using $E = hf$: (a) $E = 4.8 \times 10^{-18}\,\text{J}$, (b) $E = 3.2 \times 10^{-15}\,\text{J}$
19 Work function = $8.0 \times 10^{-19}\,\text{J} = 5.0\,\text{eV}$
20 (a) A permitted orbital is an energy level at which an electron can exist in an atom. (b) A quantum jump describes the displacement of an electron from one energy level to another. (c) The ionisation energy is the minimum energy needed to remove an electron from its ground state in the atom.
21 Using the inverse square law: as the distance is increased four times, the intensity will fall sixteen times. $I = 0.5\,\text{Wm}^{-2}$.

Note: **bold** page numbers indicate defined terms.

A

absolute refractive index **21–22**
acoustic impedance   29
active noise control   16
amplitude   6, 12
  nodes and antinodes   16–17
  worked example   17–19
analyser   25, 26–27, 86
angle of incidence   21–23, 28–29
angle of reflection   28–29
angle of refraction   21, 22
antinodes   16–17, 63–64
antiphase   10–11, 12
A-scans, ultrasound   29–30, 68
atomic spectra   49–51

B

Bohr, Niels   45, 49

C

carrier concentration   44, 90–91
charge   32–35, 41, 43–44
charge carriers   32, 35, 43, 90–91
circuits
  electrical energy transferred to
    35–36
  parallel   33–34
  potential dividers   38–39
  potentiometers   40–41
  series   33
  worked examples   34–35,
    39–40
coherence   13–15
command terms   54–55
compressions   8

conductors
  factors affecting current flow in
    43–44
  non-metallic (ohmic)   37
  questions & answers   89–90
constructive interference   15
  questions & answers   66,
    82–83
constructive superposition   12, 14
critical angle   **23**, **24,** 72–73
current (*I*)   **32**

D

DC electricity   32
  current flow in conductors
    43–44
  electrical energy and power
    35–36
  electromotive force   41–42
  Ohm's law   37
  parallel circuits   33–35
  potential dividers   38–41
  resistivity   37–38, 44
  series circuits   33
delocalised electrons   43
destructive interference   15, 66,
    83
destructive superposition   12, 14
diffraction   **27–28**
diodes   37
displacement–time graphs   6, 10
Doppler effect   **30–31**
drift velocity   43, 44, 90–91

E

Einstein, Albert   46–47, 48
electrical energy   **35–36**
electrical power   **36**

electric current   **32**
electromagnetic waves   8–10
electromotive force   **41–42**
  questions & answers   70–71,
    83–84
electronvolt (eV)   **46**
emf (electromotive force)   41–42
emission spectrum   **49**
energy levels   49–50
exam skills   54–56
excited states   49

F

fibre optics   24
filament lamps *see* tungsten
    filament lamps
free electrons   43
frequency   6
fundamental frequency   19

G

gamma-rays   9
glass fibres   24
ground state   49–50, 77

H

harmonics   19

I

impedance, acoustic   29
infrared   9
intensity of light   51–52, 64, 88
intensity (radiation flux)   **51**
interference   14–15
  applications of   15–16
  questions & answers   66,
    82–83
internal reflection *see* total
    internal reflection